PASS YOUR (
EASY WAY

By: Craig E. "Buck," K4IA

ABOUT THE AUTHOR: "Buck," as he is known on the air, received his first amateur radio license in the mid-sixties as a young teenager. Today, he holds an Amateur Extra Class and is an active instructor and Volunteer Examiner. The Rappahannock Valley Amateur Radio Club named him Elmer (Trainer) of the Year three times.

Email: K4IA@EasyWayHamBooks.com

Published by Brigade Press
130 Caroline St. Fredericksburg, Virginia 22401

Easy Way Books by Craig Buck are available at Ham Radio Outlet stores, Amazon, and online retailers:
"Pass Your Amateur Radio Technician Class Test – The Easy Way"
"Pass Your Amateur Radio General Class Test – The Easy Way"
"How to Chase, Work & Confirm DX – The Easy Way"
"How to Get on HF – The Easy Way"
"Prepper Communications – The Easy Way"

ISBN 9781095818602

Library of Congress Control Number PENDING 1

PASS YOUR GROL TEST – THE EASY WAY

TABLE OF CONTENTS

INTRODUCTION

Other commercial radio license study books take you through the questions with four possible answers on the multiple choice test. The problem with that approach is that you must read three wrong answers for every one right answer. That's 432 wrong answers and 144 right answers for GROL Element 1 and 1,800 wrong answers and 600 right answers for Element 3. Why study 2,232 wrong answers? There is nothing to learn there.

This book is different. There are no wrong answers. **The test questions and answers are in bold print to help you focus. A colon separates the question and answer.** *Hints, to help you decipher the questions and answers, are in italic print.* Many times the question will give away the answer. Hints point that out. *Cheats, mnemonic or other tricks to help you remember, are also italic.*[1]

The questions are roughly in the order they appear in the question pool. Please excuse any tortured grammar, punctuation or syntax in the bold print questions and answers. In most cases, I am copying them word-for-word from the exam.

Please direct any corrections to the author at k4ia@EasyWayHamBooks.com. This book prints on demand and edits appear within 24 hours.

Neither the test, nor this book, teaches you how to operate, choose or maintain equipment. You are encouraged to undertake the journey of further study. We'll get you the ticket to get started.

---

[1] Sometimes, there are random questions that mean nothing to you or require a memorized answer. These cheats might not be scientific but will help you recognize the right answer. They are my way of getting back at the test designers for asking such questions.

LICENSING

The Federal Communications Commission (FCC) regulates the airwaves in the United States. Among its many duties is the licensing of operators. There are several tiers of commercial radio licenses. Here is an abbreviated summary of a few from the FCC website (FCC.gov).

Restricted Radiotelephone Operator Permit (RR)
RR holders are authorized to operate most aircraft and aeronautical ground stations. They can also operate marine radiotelephone stations aboard pleasure craft (other than those carrying more than six passengers for hire on the Great Lakes or bays or tidewaters or in the open sea) when operator licensing is required.

Qualifications:

- be a legal resident of (or otherwise eligible for employment in) the United States; and
- be able to speak and hear; and
- be able to keep at least a rough written log; and
- be familiar with provisions of applicable treaties, laws, and rules which govern the radio station you will operate.

Term of License: issued for the holder's lifetime.

How to Obtain a License: Submit FCC Form 605. There is no written test for an RR.

Marine Radio Operator Permit (MP)
MPs are required to operate radiotelephone stations aboard certain vessels that sail the Great Lakes. They are also required to operate radiotelephone stations aboard vessels of more than 300 gross tons and vessels which carry more than six passengers for hire in the open sea or any tidewater area of the United States. They are also required to operate certain

aviation radiotelephone stations and certain coast radiotelephone stations.

Qualifications:

- be a legal resident of (or otherwise eligible for employment in) the United States; and
- be able to receive and transmit spoken messages in English; and
- pass a written examination as described below under Examinations.

Term of License: Issued for the holder's lifetime.

How to Obtain a License: Submit FCC Form 605 along with the appropriate Proof of Passing Certificates (for the requisite written examination element).

Examination:
To obtain an FCC Commercial Operator License, applicants must submit, to the Commission, proof of passing Element 1 - Basic radio law and operating practice with which every maritime radio operator should be familiar. To pass, an examinee must correctly answer at least 18 out of 24 questions.

General Radiotelephone Operator License (PG)
GROL is not to be confused with the amateur radio General Class license. GROL is a commercial license and does not authorize operation on amateur radio frequencies.

A PG is required to adjust, maintain, or internally repair FCC licensed radiotelephone transmitters in the aviation, maritime, and international fixed public radio services. It conveys all of the operating authority of the MP. It is also required to operate the following:

- any compulsorily equipped ship radiotelephone station operating with more than 1500 watts of peak envelope power.

- voluntarily equipped ship and aeronautical (including aircraft) stations with more than 1000 watts of peak envelope power.

Qualifications:

- be a legal resident of (or otherwise eligible for employment in) the United States; and
- be able to receive and transmit spoken messages in English; and
- pass written examinations as described below under Examinations.

Term of License: issued for the holder's lifetime.

How to Obtain a License: Submit FCC Form 605 along with the appropriate Proof of Passing Certificates.

Examinations:
To obtain an FCC Commercial Operator License, applicants must submit, to the Commission, proof of passing Written Examination Elements:

- Element 1 - Basic radio law and operating practice with which every maritime radio operator should be familiar. To pass, an examinee must correctly answer at least 18 out of 24 questions.
- Element 3 - General Radiotelephone. Electronic fundamentals and techniques required to adjust, repair, and maintain radio transmitters and receivers. The exam consists of questions from the following categories: operating procedures, radio wave propagation, radio practice, electrical principles, circuit components, practical circuits, signals and emissions, and antennas and feed lines. To pass, an examinee must correctly answer at least 75 out of 100 questions.

THE TEST

Element 1, Marine and Aviation Radio Law, is 24 questions out of a possible pool of 144. The Element 3 test, Electronic Fundamentals and Techniques, is 100 questions out of a possible pool of 600. The pool is large, but only one of six questions in the pool will show up on your exam - one from each test-subject grouping. You will not get ten questions on one topic.

The good news is that we know the exact pool of questions and answers. The test does not change anything from the pool except the order. If the pool question asks about the effect of a 470-ohm resistor, the test will not change that to a 1000-ohm resistor and expect a different answer.

You want to learn it all, but you only get one question from each group. If there is a question or concept you cannot get, skip it or try faking with a memory trick. The chances are it will not be on your test. The minimum passing score is 75%. That is 18 out of 24 correct for Element 1 and 75 out of 100 correct for Element 3. There is no Element 2.

Here's a tip: pick a day to take the test, now. Let that be your goal. Don't procrastinate. Without a goal three weeks becomes three years, maybe never. Most important: when the day comes, take the test even if you feel you are not ready. You will be better prepared than you think.

Do you know what they call the person who graduates last in his class from medical school? "Doctor" – the same as the person who graduated first in the class. My point is: pass, and no one will know the difference.

The questions are multiple choice, so you do not have to know the answer – you have to recognize it. That is a tremendous advantage for the test-taker.

However, it is difficult to study the multiple-choice format because you can be bogged down and confused by seeing three wrong answers for every correct one. Seeing wrong answers makes it harder to recognize the right answer when you see it.

The best way to study for a multiple-choice exam is to concentrate on the correct answers. That is the focus of this book. When you take the test, the correct answer should jump out at you. The wrong answers will seem strange and unfamiliar as if in a foreign language.

You should recognize the right answer without thinking. If you do not recognize the answer, analyze the question for hints. Try to eliminate the wrong answers and then guess. There is no penalty for guessing wrong, and you greatly increase your odds by eliminating the bogus responses. They are called "distractors" for a reason.

From the FCC website: *"The FCC does not administer commercial operator license examinations. To obtain a new or upgraded FCC commercial operator license, you must pass an examination administered by a COLEM[2]. The COLEMs listed below are authorized by the FCC to administer examinations nationwide and, in some instances, at foreign sites. Contact a COLEM to obtain current location and schedule information. Upon passing the examination(s), the COLEM will issue you one or more Proof of Passing Certificates (PPCs) that you must include with your license application (unless the application is filed electronically by the COLEM)."*

You can find a list of COLEMs on the FCC.gov website. Go to FCC.gov and search "Colem." Then, contact a COLEM to learn their requirements, schedules, and fees. Some COLEMs will let you test online.

---

[2] Commercial Operator License Examination Manager

THE TEST

When you pass, you receive a Proof of Passing Certificate. Most COLEMs file your Certificate and application with the FCC. You will appear in the FCC Universal Licensing System (ULS) online database in a few days. Access the ULS through the FCC.gov website and search the ULS database for your last name, first name.

You do not have to wait for the FCC to post your license. Your Proof of Passing Certificate can be sufficient. From the FCC website: *"A person who has filed an application for a commercial operator license with the FCC, and who holds a Proof of Passing Certificate(s) indicating that he or she has passed the necessary examination(s) within the previous 365 days, is authorized to exercise the rights and privileges of the applied-for operator license during the period before the FCC acts on the application, but only for a period of 90 days from the date the application was filed, provided that the applicant has not had a commercial radio operator license revoked or suspended in the past, and is not the subject of an ongoing suspension proceeding."*

The FCC does not offer diploma certificates, but you will receive a wallet-sized card. **Every commercial operator on duty and in charge of a transmitting system must: display or have in his possession an original or photocopy of his license.** (That is your first Q&A).

HOW YOU SHOULD STUDY

You ace a multiple-choice exam by learning to recognize, not recite, the answers. You could study by pouring over the multiple-choice questions. That has been the traditional way of most classes and license manuals. The problem with that method is you have to read three wrong answers to every question. That is both frustrating and confusing. Why study the WRONG answers?

The unique "Easy Way" approach has four elements. First, you never see the wrong answers so you should recognize the right answer on the test. The wrong answers will look strange and unfamiliar.

Second, the text explains the correct answer using plain English.

Third, a summary at the end of each Subelement contains a bullet version of the questions or concepts to reinforce your memory.

Fourth, many of the questions and answers are long-winded and involved. They are full of unnecessary information that can confuse you. Sometimes, all you need do is tie two words together to recognize the answer. Often, the question contains a clue to the correct answer. Hints and cheats help you decipher the question and reveal the hidden answer.

I do not recommend you take practice exams until you feel confident you have mastered this material. The reason I recommend you wait to take practice exams is, I do not want you confused by seeing all the wrong answers. Take the practice test to build confidence but study this book. You want to recognize the right answers first.

ELEMENT 1 – MARINE AND AVIATION RADIO LAW

RULES AND REGULATIONS (Subelement A)

1 - EQUIPMENT REQUIREMENTS[3]

Part 80 is the section of the FCC regulations governing stations in maritime services.
A requirement of all marine transmitting apparatus used aboard United States vessels is: only equipment that has been certified by the FCC for Part 80 operations is authorized.

Transmitting equipment authorized for use by a station in the maritime services is: only transmitters certified by the FCC for Part 80 operations. *Too many words! Part 80 governs, and the correct answers to both questions are the only ones that mention Part 80.*

Small passenger vessels that sail 20 – 100 nautical miles from land must have: an MF[4] SSB Transceiver as additional equipment. Your typical VHF radio won't carry 100 miles. The SSB (Single Side Band) transceiver operating on medium (MF) or high frequency (HF) will.

The equipment programmed to initiate distress alerts and calls to individual stations is: a DSC controller. DSC is "Digital Selective Calling." Push a button, and a distress call goes out on VHF Channel 70.

---

[3] This is Topic 1. One question comes from each of the 24 topics.
[4] MF: 300 kHz – 3 MHz. HF: 3 MHz – 30 Mhz. VHF: 30 MHz – 300 MHz. KHz is kilohertz (Thousands of Hertz). MHz is Megahertz. (Millions). "Hertz" is another way of saying "cycles per second," a measure of frequency.

The minimum transmitter power for a medium-frequency transmitter aboard a compulsorily fitted vessel is: at least 60 watts PEP. If your vessel is required to have a medium-frequency radio, it must be at least 60 watts, Peak Envelope Power.

A shipboard transmitter using FM may not exceed: 25 watts. FM (Frequency Modulation) is used for local communications. The idea is to keep power low to reduce interference to other stations.

2 - LICENSE REQUIREMENTS

To operate a fixed-tuned ship RADAR station with external controls: no radio operator authorization is required. The GROL RADAR endorsement authorizes you to adjust, repair and maintain RADAR equipment. If the equipment is fixed-tuned, it is an appliance with no special license required.

A Marine Radio Operator Permit or higher license is required for aircraft communications: when operating on frequencies below 30 MHZ not allocated exclusively to aeronautical mobile services. *Ditch the complications. If you are going to operate on frequencies not exclusive to aeronautical mobile, you need a Marine Radio Operator Permit.*

Persons ineligible to be issued a commercial radio operator license are: individuals who are unable to send and receive correctly by telephone spoken messages in English. *Too complicated. Remember, you must be able to speak English.*

GMDSS means Global Maritime Distress and Safety System. **If a passenger ship is equipped with GMDSS: there must be two operators on board with a GMDSS license.** *GMDSS rules require two operators. Cheat: 2 SS = 2 operators.*

The minimum radio operator requirement for ships subject to the Great Lakes Radio Agreement is: a Marine Radio Operator Permit. That is the permit you get by passing Element 1.

Every commercial operator on duty and in charge of a transmitting system must: display or have in his possession an original or photocopy of his license. *Just as you carry your driver's license when you drive.*

3 - WATCHKEEPING

Radio watches for compulsory radiotelephone stations will include: VHF channel 16 continuous watch. Channel 16 - the calling and distress channel.

Outside a harbor or port, compulsory equipped cargo ships keep a continuous watch on: 2182 kHz and Ch-16. Outside a harbor or port, you listen on both MF (Medium Frequency) and VHF (Very High Frequency). *Look for the answer with two frequencies.*

All compulsory, non-GMDSS vessels must: monitor Channel 16 at all times. *Too many words. Monitor Channel 16 at all times is the common answer.*

When a watch is required on 2182 kHz, a 3-minute silent period is required at: 00 and 30 minutes past the hour. Use the silent period to listen for distress calls at the top and bottom of the hour.

A required watch on VHF Ch-16:

- **Is compulsory at all times while at sea unless the vessel is in a VTS system (VTS is a vessel traffic control service similar to air traffic control).**

- **Not required when the vessel is in an A1 sea area[5] (close to shore) provided the vessel monitors both Ch-13 and VTS channel**
- **Is always compulsory in areas A2, A3, and A4 (away from shore)**
- **All the above are true**

Some form of watch is required while at sea.

The mandatory DSC watchkeeping bands/channels are: 8 MHz HF DSC, one other HF DSC, 2MHz DSC, and VHF-70. *Look for the answer with 8MHz and VHF-70.*

4 - LOGKEEPING

The person required to make entries in a required service or maintenance log is: the operator responsible for station operation or maintenance. *It makes sense that the person responsible for operation or maintenance should keep a log.*

The person responsible for the proper maintenance of station logs is: the station licensee and the radio operator in charge of the station. *The licensee and operator keep the station log.*

During a voyage, the station logs must be kept at: the principle radiotelephone operating position.

The GMDSS log should be kept at: the GMDSS operating position. *How could you maintain the logs if they were anywhere else?*

---

[5] Sea area 1 is coast to 20 nautical miles out, Sea area 2 is 20 nM to 100 nM, Sea area 3 is High seas, Sea Area 4 is polar regions

The proper procedure to make a correction in the station log is: for the original person who made the entry to strike out the error, initial the correction and indicate the date of the correction. *The original person who made the mistake must correct it.*

When there are entries related to distress or disaster, logs should be retained for: three years, unless notified by the FCC.

When there are no entries related to distress or disaster, logs should be maintained for: two years from the date of the last entry. Three years if distress or disaster and two years otherwise.

5 - LOG ENTRIES

Radiotelephone stations required to keep logs of their transmission must include:

- **Station, date and time**
- **Name of the operator on duty**
- **Station call signs with which communication took place**
- **All of the above.**

Station logs include all these things

EPIRB tests are normally logged monthly. An EPIRB is an Emergency Position Indicating Radio Beacon. Test them, and log monthly.

Key letters or abbreviations may be used in GMDSS logbooks if their meaning is noted in the log. You want the meanings to be clear to anyone who looks at the log.

Entries related to pre-voyage and pre-departure and daily tests are required as well as a summary of all required Distress communications heard and Urgency communications affecting the station's own ship.

This question combines two of four answers. Cheat: When one possible answer is two answers, that is usuall, but not always, correct.

Logging VHF safety calls is not required. This trick question asks you which statement is false. The other choices about required equipment test, operator on-and-off watch, and daily position are all true.

6 - MISCELLANEOUS RULES AND REGULATIONS

The regulations governing the use and operation of FCC-licensed ship stations in international waters are: Part 80 of the FCC rules plus the international Radio Regulations and agreements to which the United States is a party. *Once again, look for the answer with "Part 80."*

The operator of a ship radio station may allow an unlicensed person to speak over the transmitter: when under the supervision of the licensed operator. The licensed operator is in control of the station and may allow an unlicensed person to speak.

Make application for inspection of a ship GMDSS station: to the Engineer-in-Charge of the nearest FCC District Office. These annual inspections are under the direction of the FCC.

The person in ultimate control of the ship's radio station is: the master of the ship. *The master of the ship is in charge of everything.*

The principle radiotelephone operating position must be installed: in the room or an adjoining room from which the ship is normally steered. *You want to be near the steering position to pass information.*

ELEMENT 1 – RULES AND REGULATIONS

The ships that must carry radio equipment for the safety of life at sea are: cargo ships of more than 300 tons and vessels carrying more than 12 passengers. *The other answers all look good so watch for the one with more than 12 passengers.*

SUMMARY Subelement A

TOPIC 1 – RULES AND REGULATIONS

Part 80 controls.
Offshore requires an MF SSB transceiver.
DSC, Digital Signal Calling, initiates distress calls.
Minimum power is 60 watts for medium frequency.
Maximum power for FM is 25 watts.

TOPIC 2 – LICENSE REQUIREMENTS

No license required to operate fix-tuned RADAR set.
Operating frequencies not allocated to aeronautical mobile require a Marine Radio Operator Permit.
Must be able to speak English.
GDMSS, Global Maritime Distress and Safety System requires 2 operators on board.
In the Great Lakes, minimum Marine Radio Operator Permit.
Must have in possession original or copy of license.

TOPIC 3 - WATCHKEEPING

Radio watches required on Channel 16
Outside harbor or port, watch on 2182 kHz and Ch-16.
Must monitor Ch-16 at all times.
Silent period at 00 and 30 minutes past the hour.
DSC, watchkeeping on 8 MHz and VHF-70.

TOPIC 4 - LOGKEEPING

Log maintained by person responsible for station operation or maintenance.
Keep the log at the principal operating position.
GMDSS log kept at the GMDSS operating position.
Correction by the original person who made entry.

If distress or disaster entries, keep log three years.
If not, keep two years.

TOPIC 5 – LOG ENTRIES

Log includes all thing listed in the answer.
EPIRB tests monthly.
Abbreviations OK if noted in log.
Must log pre and post departure, daily tests, Distress calls heard and Urgency affecting your ship.
Logging VHF calls not required.

TOPIC 6 – MISCELLANEOUS RULES AND REGS

Part 80 controls.
An unlicensed person may speak under supervision.
Annual inspection of GMDSS is by FCC District Office.
Ship's master is in ultimate control of radio.
Radio should be in or near where ship is steered.
Ships carrying more than 12 people must have radio.

COMMUNICATIONS PROCEDURES (Subelement B)

7 - BRIDGE-TO-BRIDGE OPERATIONS

The traffic management services operated by the US Coast Guard in designated water ways to prevent ship collisions, groundings, and environmental harm is called: Vessel Traffic Service (VTS). VTS is similar to air traffic control. *Vessel Traffic Service to manage traffic.*

A bridge-to-bridge station is: a VHF radio station on a ship's bridge used only for navigational communications. *A bridge-to-bridge station is on the ship's bridge.*

Bridge-to-bridge transmissions may be more than 1 watt when broadcasting a distress message, rounding a bend or traveling in a blind spot. *You can use higher power in all these conditions.*

Bridge-to-bridge uses VHF Channel 13 (CH-13) **It is legal to use high power on Channel 13 when:**

- **Vessel called fails to respond**
- **You are in a blind situation such as rounding a bend in a river**
- **During an emergency**
- **All of these**

A ship using VHF bridge-to-bridge Channel 13 may be identified by the name of the ship in lieu of a call sign. *You may see the name of the ship but not know its call sign.*

The primary purpose of bridge-to-bridge communications is: navigation communications.

8 - OPERATING PROCEDURES – 1

The best way to minimize or prevent interference is: determine that a frequency is not in use by monitoring before transmitting. *The other answers look good but remember, do not interrupt.*

A coast station[6] may transmit a general call to a group of vessels: when announcing or proceeding the transmission of Distress, Urgency, Safety or other important messages.

When to transmit routine traffic is determined: by instructions given by the coast or government station. You can't just start blasting away. Wait for instructions from the coast or government station.

A ship station which has established initial contact with another station on 2182 kHz or Ch-16: must change to an authorized working frequency for the transmission of messages. Those are hailing or calling frequencies. Move off to pass your traffic.

A coast station notifies a ship that it has messages: by transmitting lists of call signs for which they have traffic. *Listen for the list.*

The priority of communications is: Distress, Urgency, then Safety.

---

[6] A coast station is an on-shore maritime radio station which may monitor radio distress frequencies and relays ship-to-ship and ship-to-land communications .

9 - OPERATING PROCEDURES - 2

A ship or aircraft station may interfere with a public coast station: in cases of distress.
Distress traffic is of the highest priority.

A station using telephony emissions would identify: at the beginning and end of each transmission and at 15-minute intervals.
"Transmission" means conversation.

When using an SSB station on 2182 kHz or VHF-FM on Channel 16,

- **The preliminary call must not exceed 30 seconds**
- **If no contact made, you must wait 2 minutes before repeating the call**
- **Once contact is established, you must switch to a working frequency.**
- **All of these.**

Recognize these are hailing and distress frequencies. You can't tie them up.

Before making a transmission, the station operator, except for distress calls, must: determine that the frequency is not in use by monitoring. Don't interrupt. Listen first. Distress calls always have priority, and they can interrupt.

A ship station using radiotelephony would normally call a coast station: on the appropriate ship-to-shore frequency of the coast station.
Don't tie up the hailing channels. Use the specific channel the FCC has assigned to the coast station.

In the International Phonetic Alphabet, the letters E, M and S are: Echo, Mike Sierra.
Commercial operators use the International Phonetic Alphabet. You do not have to memorize it for the test.

A--Alfa
B--Bravo
C--Charlie
D--Delta
E--Echo
F--Foxtrot
G--Golf
H--Hotel
I--India
J--Juliett
K--Kilo
L--Lima
M--Mike
N--November
O--Oscar
P--Papa
Q--Quebec
R--Romeo
S--Sierra
T--Tango
U--Uniform
V--Victor
W--Whiskey
X--X-ray
Y--Yankee
Z--Zulu

10 - DISTRESS COMMUNICATIONS

The information included in a Distress message is:

- **Name of vessel**
- **Location**
- **Type of distress and specific help requested**
- **All of the above**

The highest priority communications from ships at sea are: Distress then Urgency and Safety. *Distress is always the highest priority.*

A Distress communication: indicates the sender is threatened by grave and imminent danger and requests immediate assistance. *Distress is "grave and imminent danger."*

The order of priority of radiotelephony messages in the maritime service is: Distress, followed by Urgency and Safety and all other communications. *Distress is first again.*

Radiotelephone Distress calls and messages consist of:

- **MAYDAY spoken three times, followed by the name of the vessel and the call sign in phonetics spoken three times**

- **Particulars of position, latitude, and longitude, and other information which might facilitate rescue such as the length, color and type of vessel and the number of persons on board**
- **Nature of distress and kind of assistance required**
- **All of the above**

Rescuers would need all this information.

Distress traffic is: messages relative to immediate assistance required when threatened by grave or imminent danger such as life and safety of persons on board or man overboard. *Distress is "grave and imminent danger."*

11- URGENCY AND SAFETY COMMUNICATIONS

A typical Urgency transmission is: a request for medical assistance that does not rise to the level of a Distress or a critical weather transmission higher than Safety. *Urgency is one-step below Distress but more than a general Safety call.*

The internationally recognized Urgency signal is: PAN-PAN spoken three times before the Urgency call. PAN-PAN alerts the listeners that an Urgency message follows.

An Urgency signal shall be sent: only on the authority of either the master of the ship or the person responsible for the mobile station. *Normally the master is in charge, but he might be busy dealing with the Urgency, so the radio operator has authority.*

An Urgency signal has lower priority than: Distress.

A Safety transmission is: one which indicates a station is preparing to transmit an important navigation or weather warning.

The safety signal call word spoken three times followed by the station call letters spoken three times to announce a storm warning, danger to navigation or special aid to navigation is: Security (pronounced Say-Curitay). *The safety call word is *not* "Safety."*

12 - GMDSS

GMDSS stands for Global Marine Distress Signal Safety System. It coordinates satellite and shore stations for emergency communications. **The fundamental concept of the GMDSS is: to automate and improve emergency communications in the maritime industry.** *Look for the answer with "emergency communications."*

The primary purpose of the GMDSS is: to automate and improve emergency communications for the world's shipping industry. *Look for the answer with "emergency communications."*

The basic concept of GMDSS is:

- **Shoreside authorities and vessels can assist in a coordinated SAR[7] operation with minimum delay**
- **Search and rescue authorities ashore can be alerted to a Distress situation**
- **Shipping in the immediate vicinity will be rapidly alerted**
- **All of these.**

All the answers relate to emergency communications.

---

[7] Search And Rescue

GMDSS is primarily a system based on: the linking of search and rescue authorities ashore with shipping in the immediate vicinity of a ship in Distress. *You want to link to the immediate vicinity of a ship in Distress.*

The responsibility of vessels under GMDSS is: every ship is able to perform these communications functions that are essential for the Safety of the ship itself and other ships. *"Every ship" is responsible for aiding a vessel in Distress.*

GMDSS is required for SOLAS Convention ships of 300 gross tonnage or more. *The big boys. The International Convention for the Safety of Life at Sea (SOLAS) is an international treaty that sets safety standards for merchant ships.*

SUMMARY Subelement B

TOPIC 7 – BRIDGE-TO-BRIDGE OPERATIONS

VTS, Vessel Traffic Service prevents collisions.
Bridge-to-bridge used only for navigation communications.
B-to-B may use more than 1 watt in Distress or in a blind spot or if a vessel fails to respond. (All of the above)
B-to-B may use the vessel's name in lieu of call sign.
Primary purpose of B-to-B is navigation communications.

TOPIC 8 – OPERATING PROCEDURES -1

Prevent interference by listening before transmitting.
Coast station may transmit messages to group of vessels for Distress, Urgency or Safety or important.
Transmit routine traffic at direction of coast station.
After establishing contact, move off the calling and distress frequencies, 2182 kHz or Ch-16.
Coast station notifies of messages by transmitting list.
Communications priority is Distress, Urgency, Safety

TOPIC 9 – OPERATING PROCEDURES - 2

May interfere in case of Distress.
Identify at the beginning and end and every 15 mins.
On 2182 kHz and Ch-16, do not exceed 30 seconds for first call, wait 2 minutes for second call, move when contact established.
Before transmitting, monitor.
Call a coast station on the appropriate ship-to-shore frequency.
International phonetic "E M S" is Echo, Mike, Sierra.

TOPIC 10 – DISTRESS COMMUNICATIONS

Distress message is name of vessel, location, type of distress and help requested. (All of the above).
Highest priority is Distress then, Urgency and Safety.
Distress is grave and imminent danger.
Distress call is MAYDAY, three times.

TOPIC 11 – URGENCY AND SAFETY COMMUNICATIONS

Urgency might be medical assistance not rising to distress.
Urgency call is PAN-PAN three times.
Urgency call only on authority of ship's master or person responsible for mobile station.
Urgency has lower priority than Distress.
Safety transmission is navigation or weather warning.
Safety call is SECURITY, three times.

TOPIC 12 - GMDSS

GMDSS (Global Marine Distress Signal Safety System) purpose is to automate and improve emergency communications.
GMDSS can coordinate SAR, alert authorities, alert nearby shipping.
GMDSS links with shipping in immediate vicinity.
Every ship should be able to perform.
GMDSS required for SOLAS vessels over 300 tons

EQUIPMENT OPERATIONS
(Subelement C)

13 - VHF EQUIPMENT CONTROLS

The purpose of the INT-USA control settings on a VHF is: to change certain International Duplex channel assignments to simplex in the US for VTS and other purposes. *INT-USA selects either International or USA frequency assignments.*

Use the minimum power required to make reliable communication. That reduces interference. Often you are communicating with another vessel very close by. **VHF ship station transmitters must have the capability of reducing carrier power to: 1 watt.**

The Dual Watch function is used to: listen on any selected channel while periodically monitoring Ch-16. *Ch-16 is the distress calling channel. Dual watch lets you go about your business on another channel while the radio checks on Ch-16 periodically.*

Squelch is the circuit that silences background noise when there is no signal. **The correct setting for the manual adjustment of the squelch control is: the minimal level necessary to barely suppress any background noise.** If you turn squelch level too low, you will hear noise. Too high, and you won't hear a weak signal.

The "Scan" function is used to: sequentially scan all or selected channels. Dual Watch is only two channels. Scan runs through a list of channels you have programmed for checking.

In simplex mode, the transmit and receive functions are on the same frequency. In duplex mode, transmit and receive are on different frequencies. Duplex is usually used in conjunction with a repeater, which receives a signal and simultaneously re-broadcasts it from a higher antenna with higher power. A repeater can't transmit and receive on the same frequency or it would interfere with itself.

VHF Distress, Urgency and Safety communications (and VTS traffic calls) must be performed in Simplex operating mode: to ensure vessels not directly participating in the communications can hear both sides of the radio exchange. *In DUS situations, you want everyone to hear.*

14 - VHF CHANNEL SELECTION

The channel which VHF-FM equipped vessels must monitor at all times when the vessel is at sea is: Channel 16. *Monitor the distress calling channel.*

The aircraft frequency and emission used only for digital selective calling are: 121.500 MHz – A3E. The "A" stands for AM, amplitude modulation. *Cheat: If you want to remember this one, think A as in aircraft. Only one answer has an A (A3E).*

The VHF channel used only for digital selective calling is: Channel 70. Digital Selective Calling (DSC) sends digital codes that set off alarms on equipped receivers.

The channel used for bridge-to-bridge watch is: VHF-FM Channel 13. We've had that question before.

The channel most likely to be used for routine ship-to-ship voice traffic is: Channel 8. Every marine channel has an assigned purpose. Reserve Channel 8 for commercial ships to communicate routine traffic.

The channel used to place a call to a shore telephone is: Ch-28.

VHF Channel	Use
8	Routine ship-to-ship
13	Bridge-to-bridge
16	Distress and hailing
28	Telephone
70	Digital Selective Calling

15 - MF-HF EQUIPMENT CONTROLS

The modes to select to receive vessel traffic lists from high seas shore stations are: SSB and FEC. FEC is forward error correction, a method of checking the accuracy of data transmission. *The answer here is voice and data. SSB and FEC are the two modes.*

All MF-HF[8] Distress, Urgency and Safety communications must take place on the 6 assigned frequencies and in the simplex mode: to maximize the chances for other non-GMDSS vessels to receive those communications and to maximize the chances for GMDSS vessels to receive those communications following the transmission of a DSC call of the correct priority. *Too many words! It makes sense to stick to given frequencies so you will be heard. This is another question where two of the four choices are correct. Cheat: If you see a two-out-four option, it is often correct.*

To set up the MF/HF transceiver for a voice call to a coast station: select the J3E[9] mode for proper voice operation. *Set up a voice call for proper voice operation. You don't need to memorize J3E.*

---

[8] MF is 300 kHz to 3 MHz. HF is 3 – 30 MHz. VHF is 30 MHz – 300 MHz.

[9] J is single sideband, 3 is one channel, E is voice.

MF/HF power levels should be set to: the lowest level necessary for effective communications. This rule follows all services and modes. Minimize power to minimize interference.

To set up the MF/HF transceiver for a TELEX call to a coast station: select F1B mode or J2B mode depending on the equipment manufacturer. TELEX is teletype. The last "B" indicates "electronic telegraphy." It can be sent either by FM (F1B) or single sideband (J2B) depending on the equipment manufacturer. *Cheat: Answer with the 2 Bs: F1B J2B*

The purpose of the Receiver Incremental Tuning (RIT) or "Clarifier" control is: to "fine-tune" control of the receive frequency. *It is the "receiver" incremental tuning so it fine tunes the receive frequency.*

16 - MF-HF FREQUENCY AND EMISSION SELECTION

When using radiotelephony, a vessel would normally call another ship station on: 2182 kHz or Ch-16 unless you know the called vessel maintains a simultaneous watch on another intership working frequency. *Too many words. Call on the calling channel, 2182 kHz or Ch-16.*

The MF radiotelephony calling and Distress frequency is: 2182 kHz.

For general communication purposes, paired frequencies are: normally used with public coast stations. Paired frequencies are different frequencies for transmit and receive (Duplex). Public coast stations connect into the telephone network. Telephone is Duplex as you can hear and talk at the same time, so telephone requires different paired frequencies.

The emission to use when operating on MF distress and calling voice frequency is: J3E, single sideband telephony. *Asking for a voice frequency and SSB is the most effective voice mode. Look for the answer with SSB. The "E" is voice.*

High frequency ITU Channel 1212 is: the 12th channel in the 12 MHz band. Obviously, you need a chart to figure out what that is.

For general communications purposes, simplex frequencies are used between ship stations and private coast stations and between ship stations. These are private as opposed to public coast stations which use duplex. *This is another question where the answer is two of the above.*

17 - EQUIPMENT TESTS

The proper procedure for testing a radiotelephone installation is: to transmit the station's call sign followed by the word "test" on the frequency being used for the test. On-the-air testing is allowed. You don't have to use a dummy load, be out of port or get permission.

You may test a radiotelephone system on the air: at any time (except during silent periods) as necessary to assure proper operation.

When testing on 2182 kHz or Ch-16, testing should not continue for more than: 10 seconds in any 5-minute period. *Short tests do not tie up the calling and distress frequencies. Pick the shortest answer.*

Under GMDSS, a compulsory VHF-DSC radiotelephone installation must be tested: daily when at sea. *Proper operation of the system is a matter of life and death. Test daily.*

An NBDQ system is "Narrow Band Direct Printing" similar to TELEX. An ARQ is an automated repeat request. **The best way to test the MF-NBDP system is: to send an ARQ call to a Coast station and wait for the automatic exchange of answerbacks.** *You know it is working if you get an answer back.*

INMARSAT-C is a two-way store and forward data service. **The best way to test an INMARSAT-C terminal is: to compose and send a brief message to your INMARSAT-C terminal.** *Send yourself a message to test the system.*

18 - EQUIPMENT FAULTS

Under normal circumstances, if the transmitter aboard your ship is operating off-frequency overmodulating or distorting: stop transmitting.

If a ship radio transmitter signal becomes distorted: cease operations. *If the radio is not working properly, stop using it until you have fixed the problem.*

An indication of proper operation of an SSB transmitter rated at 60 watts would be: in SITOR (Simplex Teletype Over Radio) communications, the power meter can be seen fluctuating regularly from zero to the 60 watt reading. In SSB mode, there is no output without input. The other answers all have a steady output with no input. Teletype pulses and fluctuates the output.

An indication of a malfunction on a GMDSS station with a 24 VDC battery system would be: a constant 30 volt reading on the GMDSS console voltmeter. *A constant 30 volt reading on a 24-volt system is a problem. The charging voltage is too high. Disconnect immediately to prevent damage.*

If your antenna tuner becomes totally inoperative, to obtain operation on both the 8 MHz and 22 MHz bands bypass the antenna tuner and use a straight wire or whip antenna about 30 feet long. This length is close to resonance for those bands, and your transceiver will output power.

A symptom of malfunction in a 2182 kHz radiotelephone system that must be reported to the Master and logged is: no power output when speaking into the microphone. *"Reported and logged" is serious. No output is serious. Your radio is not transmitting.*

SUMMARY Subelement C

TOPIC 13 – VHF EQUIPMENT CONTROLS

Purpose of INT-USA setting is to change from international to US settings.
VHF transmitter must be able to reduce to 1 watt.
Dual watch periodically monitors Ch-16.
Adjust squelch to the minimum needed level to suppress noise.
Scan function scans a group of channels.
VHF Distress, Urgency and Safety calls must be made in simplex so all can hear.

TOPIC 14 – VHF CHANNEL SELECTION

Must monitor Ch-16 at all times.
Aircraft frequency and mode is 121.500 MHz A3E.
Digital selective calling on Ch-70.
Bridge-to-bridge on Ch-13.
Routine voice traffic on Ch-8.

TOPIC 15 – MF-HF EQUIPMENT CONTROLS

Vessel traffic lists are on SSB and FEC.
Distress, Urgent, Safety traffic must take place on designated frequencies.
Voice call to a coast station uses voice mode.

MF/HF power, use lowest level necessary for effective communications.
TELEX is on F1B or J2B mode.
RIT control is fine-tuning the receiver.

TOPIC 16 – MF-HF FREQUENCY AND EMISSION SELECTION
Use the calling channels unless you know the other vessel is listening elsewhere.
MF, medium frequency, distress frequency 2182 kHz.
Paired frequencies used with public coast stations.
MF Distress and calling is J3E mode, single sideband.
ITU channel 1212 is 12th channel in 12 MHZ band.
General communications use simplex.

TOPIC 17 – EQUIPMENT TESTS
Test by saying "test" and call sign.
May test any time except silent periods.
Test on 2182 kHz or Ch-16 for not more than 10 seconds in a five-minute period.
GMDSS: test daily when at sea.
Test MF-NBDP by sending message to coast station and wait for an answer back.
Test IMSARSAT-C terminal by sending yourself a message.

TOPIC 18 – EQUIPMENT FAULTS
If transmitter is malfunctioning, stop transmitting.
If transmitter is distorting, stop transmitting.
In SITOR operation, power meter fluctuates between 0 and 60 watts.
30 volt reading on a 24-volt system is malfunction.
To work on 8MHz and 22 MHz, use a 30 foot wire.
Report and log if no power output when speaking into microphone.

OTHER EQUIPMENT (Subelement D)

19 - ANTENNAS

A VHF telephony, coast maritime or ship antenna: must be vertically polarized. If one antenna is vertical and another horizontal, as much as 100 times the power transfer can be lost. *It is easier to mount antennas vertically on a ship, so everyone has to be vertical.*

The antenna requirement of a radiotelephone installation aboard a passenger vessel is: the antenna must be vertically polarized and as non-directional and efficient as is practicable for the transmission and reception of ground waves over salt water. *Too much information! The antenna must be vertical.*

The most common type of antenna for GMDSS VHF is: "none of the above.". *Trick question! "Vertical" is not a possible answer, so the correct answer is "None of the above."*

The purpose of an antenna tuner is: to alter the electrical characteristics of the antenna to match the frequency in use. *The tuner alters the electrical characteristics. It doesn't change the length or make the antenna "look" different.*

The advantage of a vertical whip over a long wire is: it radiates equally well in all directions. A long wire tends to radiate off the ends, and that is very unreliable when the boat is under way or turning.

A vertical whip antenna has a radiation pattern best described by: a circle. The old saying is, "A vertical radiates equally poorly in all directions."

20 - POWER SOURCES

For a small passenger vessel inspection, reserve power batteries must be tested: at intervals not exceeding 12 months or during the inspection.

The characteristics of the Reserve Source of Energy under GMDSS are: they must be independent of the ship's electrical system when the RSE (reserve) is needed to supply power to the GMDSS equipment. *You don't want your emergency backup power to fail when the rest of the system does. It must be independent.*

A source that provides power to radio installations for the purpose of conducting Distress and Safety communications when the vessel's main and emergency generators cannot is called: the "Reserve Source of Energy." *Ditch the complicated question and recognize a Reserve Source of Energy.*

In the event of failure of the main and emergency sources of electrical power, the term for the source required to supply the GMDSS console with power for conducting distress and other radio communications is: the reserve source of energy. *Same question as above.*

The requirement for emergency and reserve power in GMDSS radio installations is all newly constructed ships under GMDSS must have both emergency and reserve power sources for radio communications. *Got the point?*

The term "reserve source of energy" means: the supply of electrical energy sufficient to operate the radio installations for the purpose of conducting Distress and Safety communications in the event of failure of the ship's main and emergency sources of electrical power.

21 - EPIRBS

An EPIRB is: a battery-operated emergency position-indicating radio beacon that floats free of the sinking ship. An Emergency Position Indicating Radio Beacon broadcasts a distress signal. It floats free of the boat and hopefully drifts with the survivors indicating their position.

EPIRB batteries are changed: after emergency use or within the month and year replacement date printed on the EPIRB. *Watch the expiration date.* It is not one you want to miss.

If a ship sinks, the device intended to float free and is turned on automatically and transmits a distress signal is: an emergency position indicating radio beacon. *Caution: One of the answers specifies frequencies. That is wrong.*

You cancel a false EPIRB alert by: notifying the Coast Guard or rescue coordination center at once. Notify the authorities and cancel the false alarm ASAP. A general call on the radio is not enough.

A COSPAS-SARSAT system is: an international satellite-based search and rescue system. *"SAR" in SARSAT means "search and rescue."*

The advantage of a 406 MHz EPIRB is:

- **It is compatible with the COSPAS-SARSAT system, and Global Maritime Distress and Safety System (GMDSS) regulations**
- **Provides the fastest and most accurate way the Coast Guard has of locating and rescuing persons in distress**
- **Includes a digitally encoded message containing the ship's identity and nationality**
- **All of the above.**

Lots of advantages, so all are correct.

22 - SARTS

A SART is a Search and Rescue Transponder.

The frequency band used by a Search and Rescue Transponder is: 9 GHz. This is the same as the 9GHz RADAR X band.

The signal from a Search and Rescue Transponder should appear on a RADAR display: as a series of 12 equally spaced dots. They align with the source. *The dots are equally spaced because the transponder pulses are regular.*

The purpose of the SART's audible tone alarm is: it informs survivors that assistance may be nearby. *That would be a welcome sound.*

SART is: a 9 GHz transponder capable of being received by a vessel's X-Band RADAR system. *X marks the spot. Don't fall for the answer that says S-band.*

A SART begins transmitting: after it is turned on and interrogated by a 9GHz RADAR signal. The SART is in standby mode until it hears RADAR. That saves battery life.

A SART's effective range can be maximized: by holding it as high as possible. 9 GHz is strictly line-of-sight and the higher the antenna, the further it can "see."

23 - SURVIVAL CRAFT VHF

The first two questions ask which answer is *not* true.

It is not true that survival craft portable two-way VHF radios: must operate simplex on Ch-70 and at least one other channel. *Ch-70 is a digital channel and voice is prohibited.*

It is not true that survival craft portable two-way radios must: operate on Ch-13. *Ch-13 is bridge-to-bridge.*

Portable survival craft transmitters may communicate between:

- **Survival craft**
- **Survival craft and ship**
- **Survival craft and rescue unit**
- **All of the above.**

You need all the help you can get.

Equipment for radiotelephony use in survival craft stations under GMDSS must have the capability: for operation on Ch-16. *Ch-16 is the distress calling frequency. That would be very handy in a survival craft.*

Equipment for radiotelephony use in survival craft stations under GMDSS must have:

- **Operation on Ch-16**
- **Watertight**
- **Permanently affixed antenna**
- **All of these**

An SCT is a Survival Craft Transmitter. **The minimum power of the SCT is: 1 watt.**

24 - NAVTEX

NAVTEX is "navigation TELEX," an international, automated, direct-printing service for delivery of navigation and weather warnings and forecasts, as well as urgent maritime safety information to ships.

NAVTEX broadcasts are sent: with categories of messages indicated by a single letter or identifier. *The single letter identifier tells what kind of message. For instance, "A" indicates a navigational warning. NAVTEX identifies categories of messages.*

MSI (Marine Safety Information.) **MSI can be obtained by:**

- **NAVTEX**
- **SafetyNet**
- **HF NBDP**
- **All of the above**

SafetyNet is part of the Inmarsat-C system for passing messages by satellite. NBDP is Narrow Band Direct Printing. *Recognize there are several ways to get MSI.*

The primary frequency that is used exclusively for NAVTEX broadcasts is: 518 kHz. That is just below the AM band on your car radio.

To prevent the reception of unwanted broadcasts by vessels using the NAVTEX system: program the receiver to reject unwanted broadcasts. *That seems obvious.*

NAVTEX broadcasts achieve maximum range: in the middle of the night. *Just like your car radio picks up distant AM stations at night.*

The transmitting range of most NAVTEX stations is: 200-400 nautical miles.

SUMMARY Subelement D

TOPIC 19 – ANTENNAS

VHF antenna must be vertically polarized.
VHF antenna also must be efficient.
Most common type of GMDSS antenna is "none of the above."
Antenna tuner alters electrical characteristics.
Advantage of a vertical is it radiates in all directions.
Vertical whip pattern is a circle.

TOPIC 20 – POWER SOURCES

Test reserve batteries every 12 months.
Reserve Source of Energy must be independent.
Reserve Source of Energy used when main and emergency generators fail.
GMDSS equipped ships must have reserve power source.

TOPIC 21 - EPIRBS

EPIRB is battery operated emergency beacon that floats free.
EPIRB batteries changed after use or within month and year printed on the device.
Device that floats free and transmits distress call is an EPIRB.
Cancel a false EPIRB alert by notifying Coast Guard.
COSPAS-SARSAT is international search and rescue.
EPIRB is compatible with GMDSS, provides fast and accurate way for Coast Guard to locate, includes digital message identifying ship. (All of the above)

TOPIC 22 - SARTS

Search and Rescue Transponders (SART) use 9 GHZ.
SART spears on ship's RADAR as 12 dots.
SART audio tone alerts that assistance may be nearby.
SART is transponder that may be received by a vessel's X-Band RADAR.
SART begins transmitting when it is turned on and interrogated by a 9 MHz. signal.
Maximize a SART range by holding it high.

TOPIC 23 – SURVIVAL CRAFT VHF

Not true that survival craft radios must operate on Ch-70 or Ch-13.
Portable survival craft radios may communicate with anyone.
Survival craft radio must be able to monitor Ch-16.
Survival craft radio must also be watertight and have a permanent antenna.
SCT, Survival Craft Transmitter minimum power is 1 watt.

TOPIC 24 - NAVTEX

NAVTEX sends with categories of messages indicated by a single letter.
MSI, Marine Safety Information, can be sent by NAVTEX, SafetyNet or HF NBDP (all of the above).
Primary frequency for NAVTEX is 518 kHz.
To prevent receiving unwanted broadcasts, program the receiver to reject unwanted broadcasts.
NAVTEX maximum range is in the middle of night.
Transmitting range of NAVTEX is 200-400 miles.

By passing Element 1, you earn the Marine Radio Operator Permit. You must now pass Element 3 to move up to the General Radiotelephone Operator License. (GROL). There is no longer an Element 2.

ELEMENT 3 - ELECTRONIC FUNDAMENTALS & TECHNIQUES

PRINCIPLES (Subelement A)

1 - ELECTRICAL ELEMENTS

The product of the readings of an AC voltmeter and AC ammeter is called: apparent power. *The power is apparent from the readings.*

The basic unit of power is: the watt.

The term used to describe the amount of energy stored in an electrostatic field is: joules.

The device that stores electrical energy in an electrostatic field is a: capacitor. Parallel plates in the capacitor hold an electrostatic charge.

The formula to determine the inductive reactance of a coil if the frequency and coil inductance is known is: $X_L = 2\pi f L$. *Cheat: Look for 2 Pies. Only one question on the entire test has 2 pies.*

Out-of-phase power associated with inductors and capacitors is called reactive power. *Inductive and capacitive reactance makes the power (volts and amps) out of phase.*

2 - MAGNETISM

The strength of a magnetic field around a conductor is determined by: the amount of current flowing.

Current flowing through a conductor produces a magnetic field.

When induced currents produce expanding magnetic fields around conductors in a direction that opposes the original magnetic field this is known as: Lenz's law. *If you see "Lenz's law," you have found the answer.*

Opposition to the creation of magnetic lines of force in a magnetic circuit is known as: reluctance. *Opposition is reluctance.*

The term "back EMF" refers to: a voltage that opposes the applied EMF. *EMF is Electromagnetic Force (voltage). Look for the answer that mentions voltage.*

Permeability is: the ratio of magnetic flux density to the magnetizing force that produces it. *Think of it as a measure of efficiency.*

3 - MATERIALS

Galvanic corrosion is the chemical reaction of seawater and metal. **Corrosion resulting from electric current flow between dissimilar metals is called: galvanic corrosion.**

The metal usually employed as a sacrificial anode for corrosion control is: zinc bar. Zinc is very reactive, so corrosion eats at the zinc bar rather than other submerged metal on the vessel. You can replace the sacrificial zinc bar anode easily.

The relative dielectric constant for air is: 1. Other materials are compared to air.

The metal object least affected by galvanic corrosion when submerged in seawater is: a stainless steel propeller shaft. Stainless steel is not very reactive.

Skin effect is a phenomenon where: RF current flows in a thin layer of the conductor, closer to the surface as frequency increases. *Current flows on the skin of the conductor.*

The most useful insulation at UHF frequencies is: mica.

4 - RESISTANCE, CAPACITANCE AND INDUCTANCE

The formulas to calculate the total inductance of inductors in series is: $L_T = L_1 + L_2$. For inductors in series, add the values.

Good conductors with minimum resistance have: many free electrons. *Free electrons are free to pass current.*

The metals that are the best low-resistance conductors are: gold, silver, and copper. *You know copper is a good conductor so look for it in an answer.*

A bypass capacitor: removes alternating current by providing a low impedance path to ground. A capacitor passes alternating current. In this case, to ground.

The total capacitance of three capacitors in parallel is: calculated by $C_T = C_1 + C_2 + C_3$. Capacitors in parallel add.

To reduce the inductance in an antenna coil: reduce the number of turns. *Fewer turns, less inductance.*

5 – SEMI-CONDUCTORS

The most commonly-used specifications for a junction diode are: maximum forward current and PIV (peak inverse voltage). How much current it can pass and how much reverse voltage can it withstand.

The maximum forward current in a diode is limited by: the junction temperature. More current heats up the device, and heat destroys diodes.

MOSFETs are manufactured with a protective device built into their gates to protect the device from static charges and excessive voltages: the device is called a Zener diode. *Cheat: Prevent zapping with a Zener diode.*

The two basic types of junction field-effect transistors are: N-channel and P-channel. *Remember your PNP and NPN transistors.*

A common emitter amplifier has: more voltage gain than a common collector. *Cheat: Gain is emitted, not collected.*

The input impedance of a field-effect transistor compared to a bipolar transistor is: that the field effect (FET) has high input impedance and the bipolar has low input impedance. *Just remember a FET has high input impedance.*

6 - ELECTRICAL MEASUREMENTS

An AC ammeter indicates: effective (RMS) values of current). RMS (Root mean square) is a way of measuring the DC equivalent of a fluctuating alternating current.

The voltage of an AC circuit indicated on the scale of an AC voltmeter must be multiplied by:

1.414 to obtain the peak voltage value. The peak value is higher than the RMS value shown on a voltmeter.

The RMS voltage of a common household electrical circuit is: 117-VAC.

The peak voltage at a common household power outlet is: 165.5 volts. *The peak is higher than the measured RMS of 117v by a factor of 1.414. Solve: 117 x 1.414 = 165.5.*

The easiest voltage amplitude to measure viewing a pure sine-wave on an oscilloscope is: the peak-to-peak. The maximum and minimum are easy to see on the screen.

The voltage measured in an AC circuit with an AC voltmeter must be multiplied by: .9 to obtain an average. *Average is just a little less than what the RMS voltmeter tells you.*

7 - WAVEFORMS

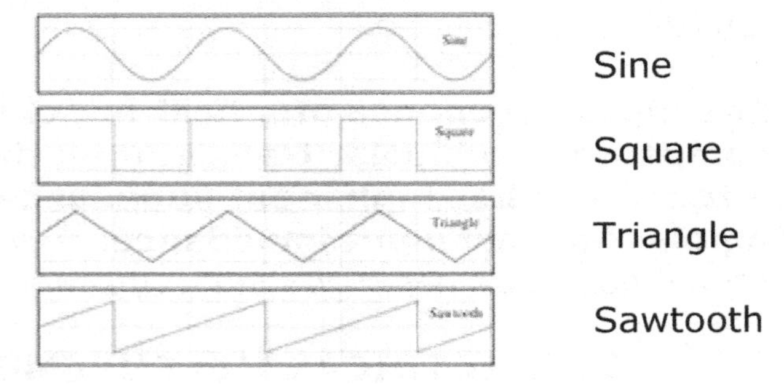

A sine wave is: a wave whose amplitude at any given instant can be represented by the projection of a point on a wheel rotating at a uniform speed.

In one complete sine wave cycle: there are 360 degrees.

A wave made up of sine waves of the fundamental frequency and all the odd harmonics is: a square wave.

The description of a square wave is: a wave that abruptly changes back and forth between two voltage levels and stays at these levels for equal amounts of time. *Sides of a square are equal.*

A wave made up of sine waves at the fundamental frequency and all the harmonics is: a sawtooth wave. Odd harmonics produce a square wave. All harmonics produce a sawtooth wave.

The wave characterized by a rise time significantly faster than the fall time (or vice versa) is: a sawtooth wave. The sawtooth wave in the graphic on the previous page has a fall time significantly faster than the rise time. The sawtooth wave is not symmetrical.

WAVE	CHARACTISTICS
SINE	Rotating wheel
Square	Abruptly changes. Odd harmonics
Sawtooth	Different rise and fall. All harmonics

8 - CONDUCTION

The terms used to identify an AC voltage that would cause the same heating in a resistor as a corresponding value of DC voltage is: Root Mean Square (RMS). RMS is a way of calculating the equivalent DC voltage.

Reactive power in a circuit that has both inductors and capacitors alternates between magnetic and electric fields and: is not dissipated. Inductors have magnetic fields, and capacitors have electric fields. Reactive power is bouncing between the two.

Halving the cross-sectional area of the conductor will: double the resistance. *A smaller hose will have more resistance than a larger one.*

The correct list in order of descending conductivity is: silver, copper, aluminum, iron and lead. *Silver and copper are the best, so pick the answer with them leading.*

To compute true power (power dissipated in the circuit) in a circuit where AC voltage and current are out of phase: multiply apparent power times the power factor. Think of the power factor as "efficiency." Mathematically, it is the cosine of the phase angle between the voltage and the current.

Assuming a power source to have a fixed value of internal resistance, maximum power will be transferred to the load when: the load impedance equals the internal impedance of the source. *Match impedances to transfer maximum power.*

SUMMARY Subelement A

TOPIC 1 – ELECTRICAL ELEMENTS

Product of AC volt and ammeter is apparent power.
Power unit is the watt.
Energy in an electrostatic field is joules.
Capacitor stores energy in electrostatic field.
Inductive reactance of coil is $2\pi f L$.
Out-of-phase power is reactive power.

TOPIC 2 - MAGNETISM

Magnetic field strength determined by current.
Current flowing produces a magnetic field.
Magnetic fields follow Lenz's law.
Opposition to magnetic lines is reluctance.
Back EMF is voltage opposition.
Permeability is ratio of magnetic flux density.

TOPIC 3 - MATERIALS

Corrosion in dissimilar metals is galvanic corrosion.
Zinc bar used as a sacrificial anode.
Dielectric constant for air is 1.
Object least affected by galvanic corrosion would be stainless steel.
Skin effect is when RF current flows on the surface.
Most useful insulation at UHF is mica.

TOPIC 4 – RESISTANCE, CAPACITANCE, INDUCTORS

Inductors in series, add the values.
Good conductors have many free electrons.
Best low-resistance conductors are gold, silver, copper.
Bypass capacitor provides low impedance path to ground.
Capacitors in parallel, add the values.
To reduce inductance in a coil, reduce number of turns.

TOPIC 5 – SEMI-CONDUCTORS

Specs for a junction diode are maximum forward current and PIV (Peak Inverse Voltage).
Maximum forward current limited by temperature.
MOSFET'S protective device is a Zener diode.
Two types of field effect transistors are N-channel and P-channel.
Common emitter has more voltage gain than common collector.
FET has high input impedance. Bi-polar has low.

TOPIC 6 – ELECTRICAL MEASUREMENTS

AC ammeter indicates RMS values.
Multiply AC voltmeter by 1.414 to get peak.
RMS voltage of house circuit is 117 VAC.
Peak voltage of house circuit is 165.5 volts.
Easiest voltage read on oscilloscope is peak-to-peak.
Multiply AC voltmeter by .9 for average.

TOPIC 7 - WAVEFORMS

Sine wave is rotating wheel.
One cycle is 360 degrees.
Sine wave and odd harmonics form a square wave.
Square wave abruptly changes for equal time.
Sine wave and all harmonics form a sawtooth wave.
Different rise and fall time is sawtooth wave.

TOPIC 8 - CONDUCTION

AC value that produces same heating as DC is RMS (Root Mean Square).
Reactive power is not dissipated.
Halving the cross-section area of a conductor, doubles the resistance.
Order of conductivity is silver, copper...
True power when voltage and current are out of phase is apparent power times the power factor.
Maximum power transferred when the load impedance matches the source.

ELECTRICAL MATH (Subelement B)

9 - OHM'S LAW – 1

Ohm's Law describes the relationship between voltage, amperage, and resistance. The math involved is multiplication or division. The formula is E=IR where E is voltage, I is amperage and R is resistance. Solve by covering the desired answer, and the formula is what is left.

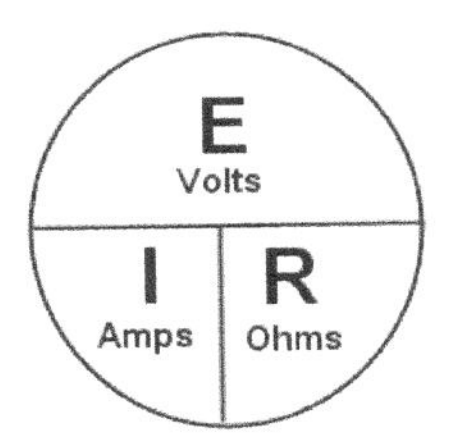

The value of series resistor needed to obtain a full-scale deflection on a 50 microamp, DC, meter with an applied voltage of 200 V DC, is: 4 megohms. R = E/I. 200/.00005 = 4,000,000. Those are microamps or millionths of an amp. Watch the decimals.

An incorrect version of Ohm's law is: I = R/E. *Use the magic circle. This one doesn't fit. Cheat: No ire allowed.*

If a current of 2 amperes flows through a 50 – ohm resistor, the voltage across the resistor is: 100 volts. *Solve: Cover the E in the magic circle, and the formula is E = I x R. and E = 2 x 50.*

If the current of 3 amperes through a resistor connected to 90 V, the resistance is: 30 ohms. *Solve: R = E/I. R = 90 / 3 = 30 ohms.*

A relay coil has 500 ohms resistance and operates on 125 mA. The value of resistance to connect in series for it to operate from 110 V DC

is: 380 ohms. *Solve: Two steps. First, what is the operating voltage of the relay? E = IR. .125 x 500 = 62.5 volts. The voltage drop in the resistor needs to be 110 – 62.5 = 47.5 volts. The resistance to get 47.5 volts is R = E/I, R = 47.5 / .125 = 380 ohms.*

10 - OHM'S LAW – 2

Power is volts times amps, P=IE and we have another magic circle. Easy as PIE.

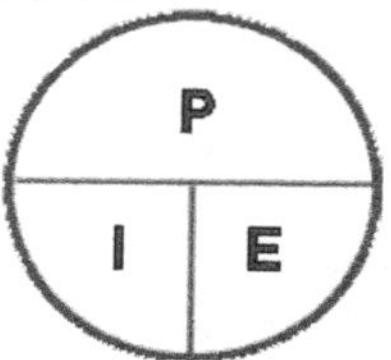

Combining the two formulas using algebra, we see that P is also equal to I^2R and E^2/R. E is also equal to the square root of the Power times Resistance.

The peak-to-peak RF voltage on the 50-ohm output of a 100-watt transmitter is: 200 volts. *Solve: First determine the voltage E = square root of 50 x 100 = 70.7. The peak RF voltage would be 1.414 times that or 99.96, and the peak-to-peak would be twice that or 200 volts.*

The maximum DC or RMS voltage that may be connected across a 20 watt, 2000 ohm resistor is: 200 volts. *Solve: P x R = E^2. 20 x 2000 = 40,000 and the square root of 40,000 is 200 volts. Cheat: The answer to both questions is 200 volts.*

A 500 ohm, 2-watt resistor and 1500-ohm 1-watt resistor are connected in parallel. The maximum voltage that can be applied across the parallel circuit without exceeding wattage ratings is: 31.6 volts. *Solve: Calculate the maximum voltage for each resistor's watt rating and the answer is the lowest. E^2 = P x R. 2 x 500 = 1000 and the square root is 31.6 for the 500-ohm. 1 x*

1500 = 1500 for the 1500-ohm. The square root is 38.7. 31.6 volts is all you can put through this circuit.

In figure 3B1, the voltage drop across R_1 is: 5 volts. *Solve: The total resistance is 600 ohms and the voltage is 10 volts. There is 10 volts drop across the entire circuit and R_1 is half the total resistance, so its drop is 5 volts. Ignore D_1 as it does not conduct in this configuration.*

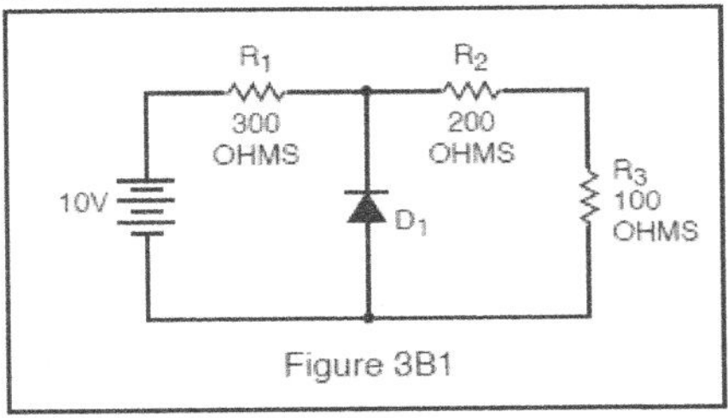

Figure 3B1

In figure 3B2, the voltage drop across R_1 is: 9 volts. *Solve: The Zener diode drops 3 volts, so the answer is 12 – 3 = 9.*

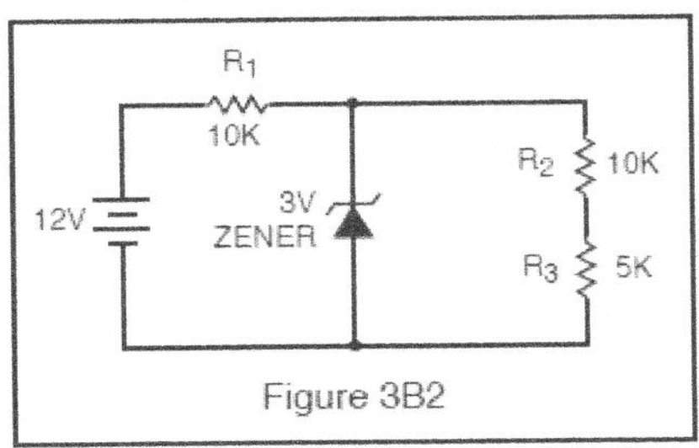

Figure 3B2

The maximum rated current-carrying capacity of a resistor marked "2000 ohms, 200 watts " is: .316 amps. *Solve: $I^2 = P / R$. $P / R = .1$ and the square root of .1 is .316 amperes. Cheat: 200 watts seems like a lot, but the answer is the smallest of the choices.*

11 - FREQUENCY

The most the actual transmitter frequency could differ from a reading of 462,100,000 hertz on a frequency counter with a time base accuracy of +/- .01 ppm is: 46.21 Hz. *Solve: .01 ppm is .01 parts per million. We have 462 million and .01 times that is 46.1*

The most the actual transmitter frequency could differ from a reading of 156,520,000 hertz on a frequency counter with a time base accuracy of +/- 1.0 ppm is: 156.52 Hz. *Solve: 1 part per million out of 156 million is 156.*

The most the actual transmitter frequency could differ from a reading of 156,520,000 hertz on a frequency counter with a time base accuracy of +/- 10 ppm is: 1565 Hz. *Solve: 10 parts per million out of 156.5 million is 1565.*

The most the actual transmitter frequency could differ from a reading of 462,100,000 hertz on a frequency counter with a time base accuracy of +/- 1.0 ppm is: 462.1 Hz. *Solve: 1 part per million on 462.1 million is 462.1.*

The second harmonic of a 380 kHz frequency is: 760 kHz. *Solve: The second harmonic is two times the fundamental. Multiply by 2.*

The second harmonic of SSB frequency 4146 kHz is: 8292 kHz. *Solve: Multiply by 2, same as above.*

12 - WAVEFORMS

At pi/3 radians, the amplitude of a sine-wave having a peak value of 5 volts is: +4.3 volts.
Solve: You could use trig or common sense.
A radian is the part of a circle where the arc length equals the radius. That would be 180 degrees. Pi/3 is

60 degrees, where the voltage is on the plus side. The voltage will be less than, but close to, peak, and it will be positive.

At 150 degrees, the amplitude of a sine-wave having a peak value of 5 volts is: +2.5 volts. *Solve: From zero to 180 degrees, the voltage is still on the plus side but barely. The peak is 5 and we are only 30 degrees positive so the voltage much less.*

At 240 degrees, the amplitude of a sine-wave having a peak value of 5 volts would be: -4.3 volts. *Solve: At 240 degrees, the voltage is on the negative side, so the answer is negative. We are close to the bottom of 270 degrees, so the answer is close to the peak negative.*

This chart is not in your test booklet. It might be helpful to sketch it in your notes.

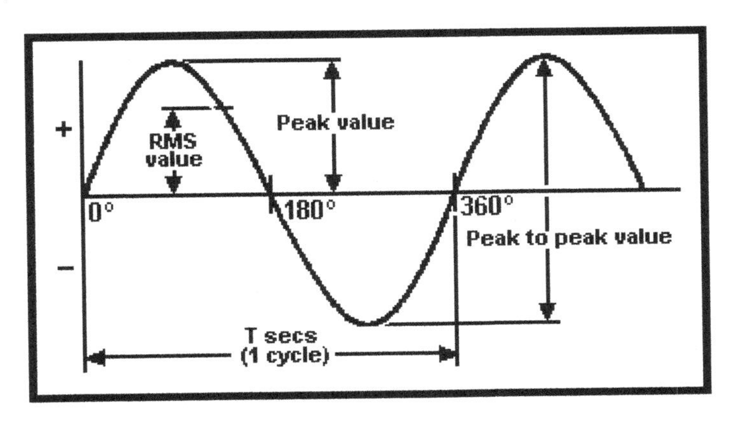

The equivalent to the root-mean-square value of an AC voltage is: the DC voltage causing the same heating in a given resistor as the RMS AC voltage of the same value. *The same heating means it has the same amount of power and is equivalent.*

The RMS value of a 340-volt peak-to-peak pure sine wave is: 120 volts. *Solve. RMS is calculated on peak voltage so first divide the peak-to-peak by 2 to get the peak. Then, multiply by .707 to arrive at the answer of 120.*

The phase relationship between the two signals in Figure 3B3 is: B is lagging A by 90 degrees. *Solve: B starts after A, so B is lagging. It is lagging by one-half the positive cycle which totals 180 degrees. One-half of 180 degrees is 90 degrees.*

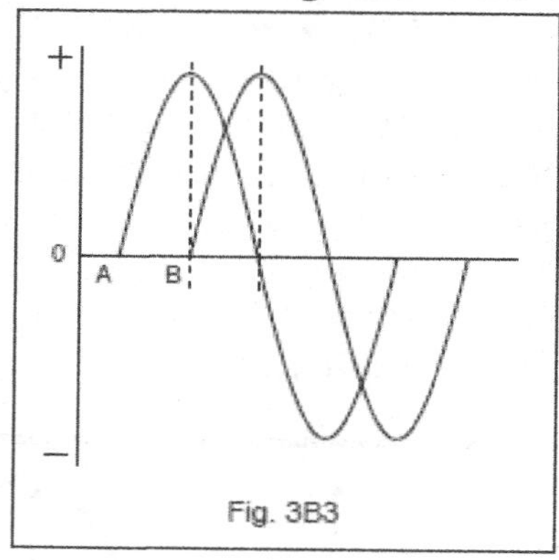

Fig. 3B3

13- POWER RELATIONSHIPS

The power factor in an R-L circuit having a 60 degree phase angle between the voltage and current is: 0.5. *Solve: The answer is the cosine of 60 degrees.* An R-L circuit is one with resistor(s) and inductor(s).

If the resistance to which a constant voltage is applied is halved: the power dissipation will double. *Solve: If E = I R, cutting the R in half will increase the I by 2. P = E I, so double the current is double the power.*

746 watts, corresponding to the lifting of 550 pounds at the rate of one-foot-per-second, is the equivalent of one horsepower. *Recognize the* *question as two different ways to express one horsepower. You will use the 746 in another question later.*

In a circuit where current and voltage are out of phase, the true power is determined: by multiplying the apparent power times the power factor. *This question was asked before.*

The power factor in an R-L circuit having a 45 degree phase angle between the voltage current is: .707. *Solve: The power factor is the cosine of the phase angle, 45 degrees.*

The power factor in an R-L circuit having a 30 degree phase angle between voltage and current is: .866. *Solve: The cosine of 30 degrees.*

You can use a scientific calculator to determine the cosine of the angle or remember:

Phase Angle	Cosine / Power Factor
30	.866
45	.707
60	.5

14 - RC TIME CONSTANTS – 1

The term for the time required for the capacitor in an RC circuit to be charged to 63.2% of the supply voltage is: one time constant An RC circuit has resistors and capacitors.

The meaning of the term "time constant of an RC circuit" is: the time required to charge a capacitor in the circuit to 63.2% of the supply voltage. *Same question as above.*

The term for the time required for the current in an RL circuit to build up to 63.2% of the

maximum value is: one time constant. *Same answer for current and voltage, 63.2%.* An RL circuit has resistors and inductors.

The term "time constant of an RL circuit" is: the time required for the <u>current</u> to build up to 63.2% of the maximum value. *This is an RL (Resistor/Inductor) circuit, so the current builds up. Voltage is near instantaneous.*

After two time constants, the capacitor in an RC circuit is charged to: 86.5% of the supply voltage. *Solve: The first time constant gets to 63.2% and the second gets to 63.2% of what is left, 36.8. 63.2 x 36.8 = 23.3% added to the charge in the first time constant of 63.2% = a total of 86.5%.*

After two time constants, the capacitor in an RC circuit is discharged: to 13.5% of the starting voltage. *Solve: This is the discharge cycle, and the question is asking what is remaining in the capacitor. The calculation is the reverse but the answer is related. The first time constant took it down to 36.8% (100- 63.2) and the second takes it down another 23.3% (36.8 x 63.2) leaving 13.5%. If the voltage goes up 86.5% in two time constants on the charge side, on the discharge side, 100-86.5% remains. Cheat: Charging for two time constants is 86.5% and discharging for two time constants leaves 13.5%*

15 - RC TIME CONSTANTS – 2

Multiply the farads by ohms to get the value of one time constant. These questions complicate the solution by first requiring that you calculate the farads and ohms of components in parallel and series.

The time constant of the circuit having two 220-microfarad capacitors and two one-megohm resistors all in parallel is 220 seconds. *Solve: The two capacitors in parallel total 440 microfarads.*

The two resistors in parallel are .5 megohms. Multiply the 440 x .5 and get 220 seconds.

The time constant of a circuit having two 100-microfarad capacitors and two 470-kilohm resistors all in series is 47 seconds. *Solve: The two capacitors in series are 50 microfarads. The two resistors are 940 kilohms. We have to get the decimals straight. 50 microfarads is .00005 farads and .00005 x 940,000 = 47 seconds.*

The time constant of a circuit having a 100-microfarad capacitor and a 470-kilohm resistor in series is 47 seconds. *Solve: 100 microfarads is .0001 farads times 470,000 = 47.*

The time constant for a circuit having two 100-microfarad capacitors and two 470-kilohm resistors all in parallel is 47 seconds. *Solve: The two capacitors total 200 microfarads (.0002) The two resistors are 235 kilohms. .0002 x 235,000 = 47.*

The time constant of a circuit having two 220-microfarad capacitors and two 1-megohm resistors all in series is 220 seconds. *Solve: The two capacitors in series are 110 microfarads (.00011), and the two resistors are 2 megohms. .00011 x 2,000,000 = 220*

By now, you either are an expert or have thrown your hands up in despair. Fear not, here is the ultimate cheat that answers these questions. If you ever have to perform such calculations in the future, you can look them up.[10] *Cheat: If the question has a 220-microfarad capacitor, the answer is 220 seconds. If the question has a 470-kilohm resistor, the answer is 47 seconds.*

---

[10] Einstein said, "Never memorize anything you can look up in a book. "

16 - IMPEDANCE NETWORKS – 1

The following questions ask about components in series. In an inductive circuit, the value of "j" is positive. In a capacitive circuit, the value is negative.

The impedance of a network composed of a 0.1-microhenry inductor in series with a 20-ohm resistor, at 30 MHz, in rectangular coordinates is: 20 +j19. *Solve/Cheat: You know the first number is the resistance (20). The second number will be positive because this circuit has an inductor. Pick the only answer with 20 and +j.*

In rectangular coordinates, the impedance of a network composed of a 0.1-microhenry inductor in series with a 30-ohm resistor, at 5 MHz is: 30 +j3. *Solve/Cheat: The first number is resistance (30). The second will be positive because it is an inductor. Pick the answer with 30 and +j.*

In rectangular coordinates, the impedance of a network composed of a 10-microhenry inductor in series with a 40-ohm resistor, at 500 MHz is: 40 +j31400. *Solve/Cheat: The first number is resistance (40). The second will be positive because it is an inductor. Pick the answer with 40 and +j.*

In rectangular coordinates, the impedance of a network composed of a 1.0-millihenry inductor in series with a 200-ohm resistor, at 30MHz is: 200 +j188. *Solve/Cheat: The first number is resistance (200). The second will be positive because it is an inductor. Pick the answer with 200 and +j.*

In rectangular coordinates, the impedance of a network composed of a .001-microfarad capacitor in series with a 400-ohm resistor, at 500 kHz is: 400 –j318. *Solve/Cheat: The first number is resistance (400). The second will be* *negative because the component is a capacitor. Pick the answer with 400 and -j.*

Now, we move to components in parallel. The computations get much more complicated, and the cheats for series circuits no longer apply.

In rectangular coordinates, the impedance of a network composed of a 0.01-microfarad capacitor in parallel with a 300-ohm resistor, at 50 MHz is 159 –j150. *Solve/Cheat: You know the answer is –j because the circuit has a capacitor. That eliminates two answers. The resistance is 159 ohms.*

17 - IMPEDANCE NETWORKS – 2

This topic deals with polar coordinates rather than rectangular coordinates. The answers are ohms and degrees. Capacitive circuits are - degrees. Inductive circuits are + degrees. Unless you are prepared to engage in some very complicated trigonometry, we need some tricks to get through this section.

The impedance of a network composed of a 100-picofarad capacitor in parallel with a 4000-ohm resistor, at 500KHz in polar coordinates is: 2490 ohms, /-51.5 degrees. *Solve/Cheat: The circuit has a capacitor, so the answer is negative. That eliminates two answers. The components are in parallel, so the total reactance is less than the 4000-ohm resistor.*

In polar coordinates, the impedance of a network composed of a 100-ohm reactance inductor in series with a 100-ohm resistor is: 141 ohms, /45 degrees. *Solve/Cheat: If the reactance and resistance are equal, the phase angle is 45 degrees.*

In polar coordinates, the impedance of a network composed of a 400-ohm-reactance capacitor in series with a 300-ohm resistor is: 500 ohms, /-53.1 degrees. *Solve/Cheat: The answer must be negative because of the capacitor. The reactances are in series, so the total is the higher number of the answers, 500 ohms.*

In polar coordinates, the impedance of a network composed of a 300-ohm-reactance capacitor, a 600-ohm reactance inductor and a 400-ohm resistor, all connected in series is: 500 ohms, /37 degrees. *Solve/Cheat: The reactance is net inductive, so the answer must be positive. The reactances are in series, so the total is the higher number of the answers, 500 ohms.*

In polar coordinates, the impedance of a network comprised of a 400-ohm-reactance inductor in parallel with a 300-ohm resistor is: 240 ohms /36.9 degrees. *Solve/Cheat: The circuit is inductive, so the answer is positive. The two reactances are in parallel, so the total is less than either one.*

Using a polar coordinate system, the visual representation of a voltage in a sinewave circuit: shows the magnitude and phase angle. A polar coordinate plot shows magnitude and phase angle.

18 - CALCULATIONS

The magnitude of the impedance of a series AC circuit having a resistance of 6 ohms, an inductive reactance of 17 ohms and zero capacitive reactance is: 18 ohms. *Solve: The formula is* $R^2=I^2+C^2$. $6^2=36$. $17^2=289$. *36+289 = 325 and the square root of 325 is 18.*

A 1-watt, 10-volt Zener diode with the following characteristics: I_{min} **= 5 mA;** I_{max} **= 95 mA; and Z**

= 8 ohms is to be used as part of a voltage regulator in a 20-V power supply. The size of the current-limiting resistor to set bias to the midpoint of its operating range would be: 200 ohms. *Solve: A lot of extraneous information in the question. The voltage is 10 volts; the current is between 5 mA and 95 mA, call the midpoint 50 mA. Ohms law, R = E/I, says the resistor should be: 10 / .05 = 200 ohms.*

Given a power supply with a no-load voltage of 12 volts and a full load voltage of 10 volts, the percentage of voltage regulation is: 20%. *Solve: The variance is 2 volts out of 10 volts or 20%. Figure from the full-load voltage.*

Given a power supply with a full load voltage of 200 volts and a regulation of 25%, the no-load voltage is: 250 volts. *Solve: 25% of the full load voltage is 50 volts so the no-load voltage must be 200 + 50 = 250.*

The turns ratio on a transformer to match a source impedance of 500 ohms to a load of 10 ohms is: 7.1 to 1. *Solve: Impedance turns go by the square root of the impedance ratio. Here the impedance ratio is 50. The square root of 50 is 7.1.*

The conductance (G) of a circuit if 6 amperes of current flows when 12 volts DC is applied is: .5 Siemens. *Solve: Conductance is the reciprocal of resistance. A reciprocal is a number divided into 1. Applying Ohms law, the resistance is 12/6 =2 ohms. The reciprocal of 2 is 1/2 = .5 Siemens.*

SUMMARY Subelement B

TOPIC 9 – OHM'S LAW – 1

R = E/I
I = R/E is incorrect.
E = IR

TOPIC 10 – OHM'S LAW – 2

Know or be able to calculate the following:
E = square root of PR
Peak = 1.414 AC volts
P x R = E^2
I^2 = P/R

TOPIC 11- FREQUENCY

+/- .01 ppm is .01 parts per million
Second harmonic is two times the fundamental frequency.

TOPIC 12 - WAVEFORMS

Pi/3 radians is 60 degrees.
Voltage is positive from 0 – 180 degrees.
Voltage is negative from 180 0 360 degrees.
RMS voltage is same heating as DC voltage.
RMS voltage is .707 times peak.

TOPIC 13 – POWER RELATIONSHIPS

Power factor is cosine of phase angle.
746 watts equals one horsepower.
True power = apparent power times power factor.

TOPIC 14 – RC TIME CONSTANTS -1

Charge of 63.2% takes one time constant.
Two time constants is 86.5% charge.
In an RL circuit, current builds up.
Two time constants of discharge leaves 13.5%

TOPIC 15 – RC TIME CONSTANTS -2

Time constant = farads times ohms.
If question has a 220 µf capacitor, answer is 220 seconds.

If question has a 470 kilohm resistor, answer is 47 seconds.

TOPIC 16 – IMPEDANCE NETWORKS – 1

Rectangular coordinates are +j for inductive, -j for capacitive.
First number is resistance.

TOPIC 17 – IMPEDANCE NETWORKS – 2

Polar coordinates show magnitude and phase angle.
+ degrees for inductive, - degrees for capacitive.
If components are parallel, resistance is less.
If components are in series, resistance is more.

TOPIC 18 – CALCULATIONS

Magnitude of impedance is R^2 = $Reactance^2$ + $Resistance^2$
Voltage regulation is drop / load voltage.
Impedance matching transformer, ratio of turns is the square root of the impedance ratio.
Siemens are a unit of conductance and the reciprocal of ohms. 1/R = Siemens.

COMPONENTS (Subelement C)

19 - PHOTOCONDUCTIVE DEVICES

When light shines on photoconductive material: the conductivity increases. *It is photoconductive, not photoresistive.*

The photoconductive effect is: the increased conductivity of an illuminated semiconductor junction.

In crystalline solids, the photoconductive effect produces a noticeable change: in the resistance of the solid.

When a photosensitive semiconductor junction is illuminated: the junction resistance decreases. It becomes more conductive.

An optoisolator is: an LED and a photosensitive device. The light-emitting diode (LED) shines on a photosensitive device and transfers signals without having an electrical connection. An optoisolator passes a signal while blocking electrical noise.

An optocoupler is: an LED and photosensitive device. Optocoupler is another word for optoisolator. It couples and isolates. They are the same thing. "Opto" for optical.

20 - CAPACITORS

The factors that determine the capacitance of a capacitor are: the distance between the plates and the dielectric constant of the material between the plates. The amount of capacitance is all about the design of the component, not the voltage applied to it.

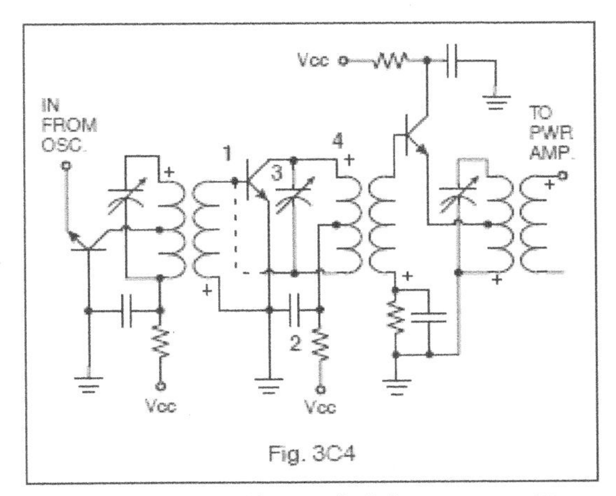

Fig. 3C4

In Figure 3C4, a small variable capacitor installed in place of the dashed line would: decrease parasitic oscillations. It provides negative feedback around the transistor.

In figure 3C4, (above) the component used to provide a signal ground is: number 2. It is a bypass capacitor and bypasses AC signal to ground.

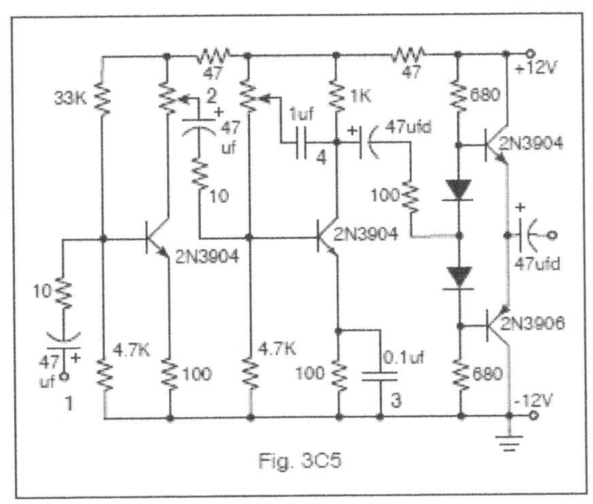

Fig. 3C5

In Figure 3C5, the capacitor used as a bypass capacitor is number: 3. It passes to ground.

In Figure 3C5, the 1μF capacitor connected to the potentiometer is used to: adjust tone.

The purpose of a coupling capacitor is: to block direct current and pass alternating current. Capacitors block DC and pass AC. A coupling capacitor couples the AC.

21 - TRANSFORMERS

A capacitor placed in series with the primary of a power transformer: improves the power factor.

A transformer used to step up its input voltage must have: more turns on its secondary than on its primary. The additional turns step up the voltage.

A transformer primary of 2250 turns connected to 120 VAC with 500 turns on the secondary develops: 26.7 volts. *Solve: Ratio of turns is 500/2250 = .222 x 120 volts = 26.7.* Fewer turns in the secondary means this transformer steps down. Note, voltage transformer ratios go by the ratio of the turns. Impedance transformer's turns go by the square root of the impedance ratio.

The ratio of output frequency to input frequency of a single-phase full-wave rectifier is: 2:1. The full-wave rectifier inverts half the wave.

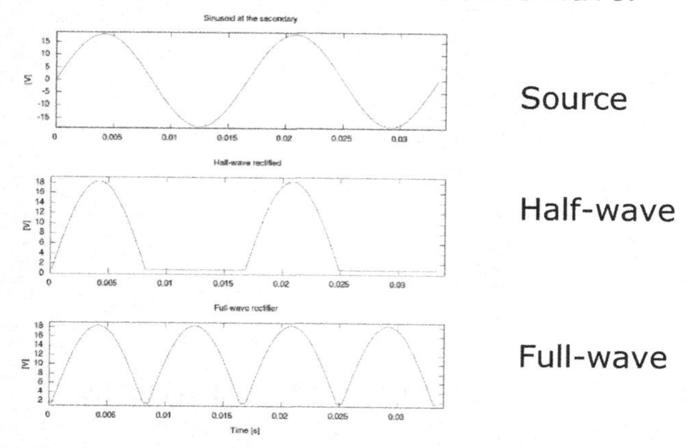

A power transformer has a single primary winding and three secondary windings producing 5.0 volts, 12.6 volts, and 150 volts. The

windings with the highest measured DC resistance will be: the 150-volt winding. It has more turns and therefore more wire and more DC resistance.

A power transformer has a primary winding of 200 turns of #24 wire and a secondary winding consisting of 500 turns of the same size. When 20 volts are applied to the primary, the secondary voltage will be: 50 volts. *Ignore the wire size. The transformation is the ratio of the turns. The secondary has 2.5 times the primary turns, and the voltage will be 2.5 times the primary.*

22 - VOLTAGE REGULATORS, ZENER DIODES

In a linear electronic voltage regulator: the conduction of a control element is varied in direct proportion to the line voltage or load current. *Vary the voltage or current by changing the conduction. It is linear, not switching.*

A switching electronic voltage regulator: switches the control device on or off with the duty cycle proportionate to the line or loan conditions. *A "switching" regulator switches.*

A Zener diode will pass current once the voltage reaches the rated amount. This prevents the voltage from going any higher. Therefore, the Zener diode works as a voltage regulator.

The device used as a stable reference voltage in a linear voltage regulator is: a Zener diode. *A stable reference voltage is provided by a voltage regulator.*

In a regulated power supply, the component most likely used to establish a reference voltage is: a Zener diode.

A three-terminal regulator contains: a voltage reference, error amplifier, sensing resistors and transistors and a pass element. *There may be three terminals but look for the answer with five elements!*

The range of voltage ratings available in Zener diodes is: 2.4 volts to 200 volts and above. *The minimum is 2.4 volts. Remember that, and there is only one answer.*

23 - SCRs, TRIACS

An SCR is a silicon controlled rectifier. SCRs can handle high voltages and currents. Back-to-back they control both sides of an AC power sine wave. Rectifiers are also called diodes.

To safely distribute the power load in a circuit, two similar SCRs might be: connected in parallel, reverse polarity. *Back-to-back.*

The three terminals of an SCR are: anode, cathode, and gate. *Cheat: Diodes (rectifiers) have an anode and cathode. Only one answer has an anode and cathode.*

Two SCRs connected back to back, but facing in opposite directions and sharing a common gate are called a: TRIAC. *Think "SCRs have three terminals, therefore TRIAC."*

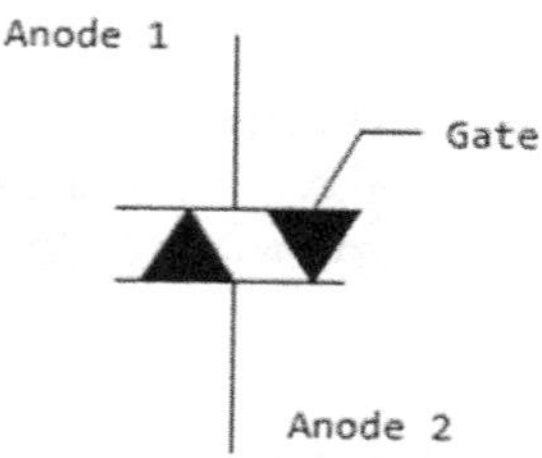

A transistor fabricated as two complimentary SCRs in parallel with a common gate terminal is: a TRIAC. *Same question.*

The three terminals of a TRIAC are: gate, anode 1 and anode 2. *Cheat: It is a diode, so it has to have an anode. Only one answer has "anode."*

The terminal called a "gate" acts to regulate the current flow through the SCR. **A circuit that might contain an SCR is: a light-dimming circuit.**

24 - DIODES

A common use for PIN diodes is: an RF switch. PIN diodes are often used in place of mechanical relays to switch from transmit to receive.

A hot carrier diode is also called a Schottky diode, named for its inventor. Hot carrier diodes are known for fast switching and low noise.

A common use of a hot-carrier diode is: VHF and UHF mixers and detectors.

The two main structural categories of semiconductor diodes are: junction and point contact. Junction and point contact describe the internal design.

A diode capable of both amplification and oscillation is: a tunnel diode.

The principal characteristic of a tunnel diode is: it has a negative resistance region. "Negative resistance" means current decreases as the voltage increases.

The type of semiconductor diode used that varies its internal capacitance as the voltage varies is

called: a varactor diode. *If capacitance varies, it is a varactor.*

25 - TRANSISTORS – 1

With regard to bipolar transistors, the term "alpha" refers to a change: in collector current with respect to emitter current. "CCEC" is a measure of the current flowing through a transistor.

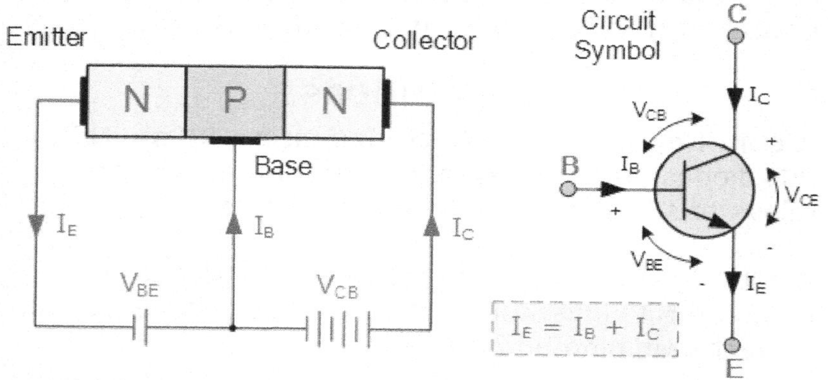

The three terminals of a bipolar transistor are: base, collector, and emitter.

With regard to bipolar transistors, the term "beta" refers to: collector current with respect to base current. It is a measure of the amplification. A small change at the base makes a big change on the collector/emitter. *Beta = base.*

The elements of a unijunction transistor are: base 1, base 2 and emitter. *"Unijunction" is one junction to two bases.* The next page pictures a unijunction transistor.

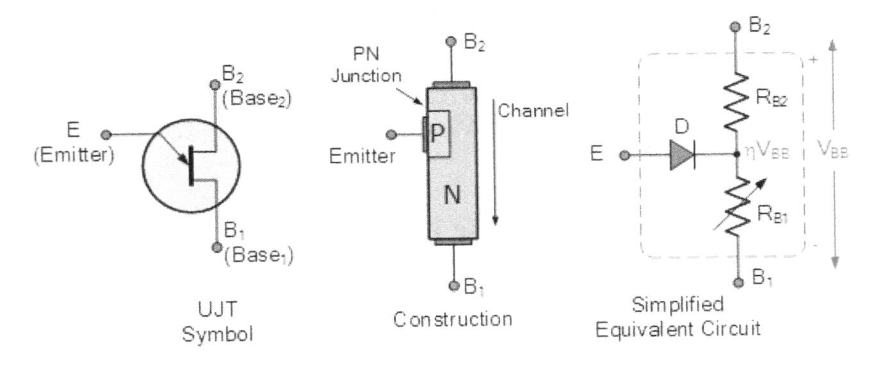

The beta cutoff frequency of a bipolar transistor is: the frequency at which emitter current gain has decreased to .707 of maximum. *Look for the answer with "decrease of maximum."*

To say a transistor is fully saturated means: the collector current is at its maximum value. *The collector can't take any more current.*

26 - TRANSISTORS – 2

A common base amplifier has more voltage gain than a common emitter or common collector. *Look for the answer with "more voltage gain."*

For a transistor to be cut off means: there is no current between emitter and collector. *Amplification has stopped.*

An emitter-follower amplifier has more current gain than common emitter or common base.

Common base	More voltage gain
Emitter-follower	More current gain.

When a transistor is operating in saturation:, the base-emitter and collector-base junctions are both forward biased. *To be in saturation, everything must be forward biased. The other answers don't have both junctions forward biased.*

For current to flow in an NPN silicon transistor's emitter-collector junction, the base must be: at least .7 volts positive with respect to the emitter. *The minimum voltage to kick-start a silicon conductor is about .7 volts.*

When an NPN transistor is operating as a Class A amplifier: the base-emitter junction is forward biased, and the collector-base junction is reverse biased. *Cheat: The minimum to remember: base-emitter is forward and the other is reverse.*

27 - LIGHT EMITTING DIODES

The bias required for an LED to produce luminescence (light) is: forward bias. *It won't conduct otherwise.*

The visible color radiated by an LED junction is determined by: the materials used to construct the device.

The approximate operating current on a light-emitting diode is: 20 mA.

The maximum current to safely illuminate an LED would be: 20 mA.[11]

An LED facing a photodiode in a light-tight enclosure is commonly known as an: optoisolator. *This question is also with the Photoconductive devices in Topic 19.*

The circuit component connected in series to protect an LED would be: a series resistor. *A series resistor is connected in series to limit current.*

[11] This is an out-dated question. LEDs do much more than that today.

28 - DEVICES

A diode junction that is forward biased is: a low impedance. *If it is forward biased, it is prone to conducting and therefore low impedance (resistance).*

CMOS stands for: "complementary metal-oxide semiconductor." *It is a semiconductor.*

FETs are field effect transistors.
The special precautions in handling FET and CMOS devices is: they are susceptible to damage from static charges. Wear a grounding strap or touch an electronic ground when working with them.

The piezoelectric effect is: mechanical vibration of a crystal by the application of a voltage. *Caution: We think of the spark generated by a cigarette lighter. Generation of electrical energy is not the answer. This is the reverse, the excitation of the crystal by applying a voltage.*

The oscillator circuit that uses a crystal is called a: Pierce. Named for its inventor George W. Pierce, the crystal determines the frequency. Colpitts oscillators use a capacitor.

An electrical relay is: a remote controlled switching device.

SUMMARY Subelement C

TOPIC 19 – PHOTOCONDUCTIVE DEVICES

Light increases conductivity.
Light reduces resistance in the solid.
Optoisolator is LED and photosensitive device.
Optocoupler is LED and photosensitive device.

ELEMENT 3 - COMPONENTS

TOPIC 20 – CAPACITORS

Capacitance determined by distance between plates and dielectric.
Capacitors can be used to: decrease parasitic oscillations, provide signal ground, bypass, adjust tone, coupling, block DC and pass alternating current. (All of the above).

TOPIC 21 – TRANSFORMERS

Capacitor in series with primary improves power factor.
Step up has more turns on secondary.
Ratio of turns determines amount of step up or down.
Full-wave rectifier output is twice input frequency.
High voltage winding will have highest resistance.

TOPIC 22 – VOLTAGE REGULATORS, ZENER DIODES

Conduction varies in proportion to line voltage in a linear regulator.
Switching regulator switches device on or off.
Stable reference voltage controlled by Zener diode.
Three terminal regulator has five elements.
Minimum Zener voltage is 2.4 volts.

TOPIC 23 – SCRs, TRIACS

Silicon controlled rectifier.
Distribute power load with back-to-back SCRs.
Three terminals are anode, cathode and gate.
Two SCRS back-to-back are a TRIAC.
Two SCRs in parallel with a common gate are TRIAC.
Three terminals of TRIAC: gate, anode 1 and anode 2.
SCR used in light-dimming circuit.

TOPIC 24 – DIODES

PIN diodes used in RF switch.
Hot-carrier diodes used in VHF/UHF mixers and detectors.
Structural categories are junction and point contact.

Tunnel diode is capable of both amplification and oscillation. Has a negative resistance region.
Varactor: voltage varies internal capacitance

TOPIC 25 – TRANSISTORS – 1

Alpha is change in collector current vs. emitter current
Three terminals of bipolar: base, collector, emitter.
Beta is collector current vs. base current.
Unijunction transistor has base 1, base 2, emitter.
Beta cutoff frequency is emitter current at .707 of max
Fully saturated means collector current at max value.

TOPIC 26 – TRANSISTORS – 2

Common base has more voltage gain.
Cut off means no current between emitter and collector.
Emitter follower has more current gain.
In saturation, junctions are both forward biased.
Base must be >.7 volts positive for current to flow
NPN in Class A has base-emitter forward biased and collector-base is reverse biased.

TOPIC 27 – LIGHT EMITTING DIODES

Bias to illuminate is forward bias.
Color determined by materials
Operating and maximum current 20 mA.
LED facing photodiode is optoisolator.
Protect an LED with a series resistor.

TOPIC 28 – DEVICES

Forward biased junction is low impedance.
CMOS is Complimentary Metal-Oxide Semiconductor.
CMOS devices subject to damage from static.
Piezoelectric effect is mechanical vibration by voltage.
Crystal oscillator is Pierce.
Electrical relay is remote controlled switching device.

CIRCUITS (Subelement D)

29 - R-L-C CIRCUITS

The approximate magnitude of the impedance of a parallel R-L-C circuit at resonance is: approximately equal to the circuit resistance. *At resonance there is no capacitive or inductive reactance, so the circuit resistance is all that is left.*

The approximate magnitude of the impedance of a series R-L-C circuit at resonance is: approximately equal to the circuit resistance. *Series or parallel, the same answer as above.*

Voltage across reactances in series could be greater than the applied voltage because of: resonance. The voltages could be higher because the inductive and capacitive reactances are returning energy back to the circuit. At resonance, they return the maximum energy.

The current flow in a series R-L-C circuit at resonance is: maximum. *This is a variation on the question before. At resonance, current flow is maximized.*

The current flow within the parallel elements in a parallel R-L-C circuit at resonance are: maximum. *Series or parallel, the answer is the same.*

The relationship between current and voltage in a resonant circuit is: the current and voltage are in phase. In a non-resonant circuit, current and voltage are out-of-phase.

30 - OP AMPS

The main advantage of using an op-amp filter over a passive LC audio filter is: op-amps exhibit gain rather than insertion loss.

The characteristics of an inverting operational amplifier (op-amp) circuit are: it has input and output signals 180 degrees out of phase. *The signals are inverted in an inverting op-amp.*

A closed loop control system is one is which the output is fed back to the input node. Therefore, the output is dependent on the feedback.

An open loop control system is one in which the output is generally only dependent upon the input and has no feedback.

Gain in a closed-loop op-amp is determined by: the op-amp's external feedback network. Op-amps rely on an external feedback network. The design of that network determines the op-amp's gain.

The external feedback network to control the gain of closed-loop op-amp is connected: from the output to the inverting input. *Op-amps are inverting. It is feedback, so it must go from the output to the inverting input.*

The op-amp circuit operated open-loop is a comparator. A comparator compares two voltages or currents and outputs a digital signal indicating which is larger. There is no feedback and the output is only dependent on comparing two inputs.

In the op-amp circuit shown in figure 3D6, the most noticeable effect if the capacitance of C were suddenly doubled is: the frequency would be lower. Op-amps rely on feedback, which is another way of saying oscillation. If the capacitance were doubled, the frequency of oscillation would be lower.

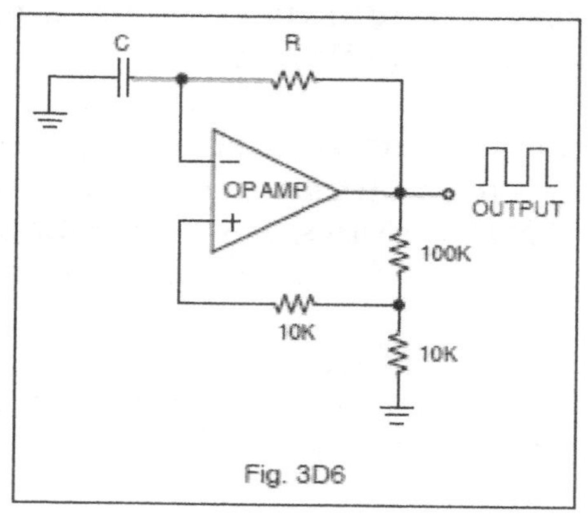

Fig. 3D6

31 - PHASE LOCKED LOOPS (PLLS); VOLTAGE CONTROLLED OSCILLATORS (VCOS); MIXERS

The frequency synthesizer circuit that uses a phase comparator, look-up table digital-to-analog converter and a low-pass antialias filter is: a direct digital synthesizer. *The digital-to-analog converter makes it a digital synthesizer.*

In a direct digital synthesizer, the unwanted components on its output are: spurs at discrete frequencies. *Digital-to-analog converters generate spurs.*

A circuit compares the output of the voltage-controlled oscillator (VCO) to a frequency standard and produces an error voltage that is then used to adjust the capacitance of a varactor diode used to control frequency in that same VCO is: phase-locked loop. *The error voltage is fed back to adjust. It sounds like a loop. Phase-locked loop.*

The definition of a phase-locked loop (PLL) circuit is a servo loop consisting of a phase

detector, low-pass filter, and a voltage-controlled oscillator. *It is "phase-locked" so it must have a phase detector.*

The spectral impurity components that might be generated by a phase-locked-loop synthesizer is: broadband noise. It is also called "phase noise."

Digital-to-analog generates spurs. Phase-locked loop generates broadband noise.

RF input to a mixer is 200 MHz and the local oscillator frequency is 150 MHZ. The output at the IF output prior to any filtering would be: 50, 150, 200 and 350 MHz. *Before filtering, you would have the original frequencies, and the sum and difference frequencies.*

32 - SCHEMATICS

Given the combined DC input voltages, the output voltage from the circuit shown in Figure 3D7 would be -5.5 V. *Solve: The total voltage input is 550 mV. That is reduced by a factor of 10K and amplified by a factor of 100K in the op-amp, so the result is 10 times the 550 mV or 5.5 volts. The symbol is for an op-amp, so the result is inverted and the answer is -5.5 V.*

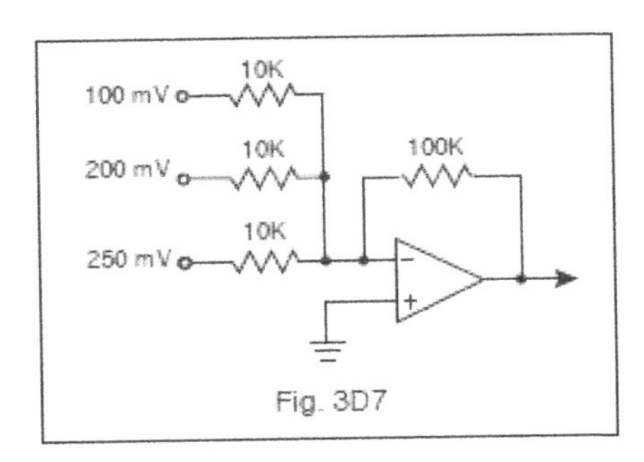

Fig. 3D7

The lamps that would be lit in Figure 3D8 are: 2, 3, 4, 7 and 8. *Solve: First, we have to accept the fiction that electricity goes from positive to negative because that is the way the symbols for diodes make sense. The arrow points the way. If you follow these around, you will see a path through 2, 3, 4, 7 and 8.*

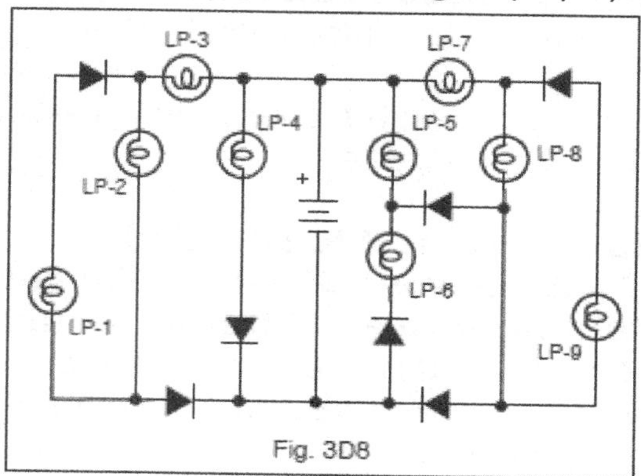

Fig. 3D8

If an amplifier input signal coupling capacitor fails open: no amplification will occur with DC within the circuit measuring normal. *If the capacitor fails open, no signal will pass to the amplifier. The fact the DC reads normal is information you don't need to get the correct answer.*

In Figure 3D9, the problem with this regulated power supply is: there is no problem. *That makes for an easy answer.*

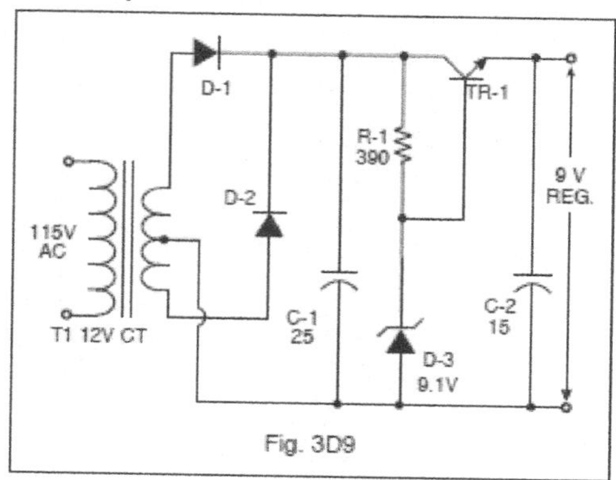

Fig. 3D9

In Figure 3D10, with a square wave input, the output would be: #3. *The capacitor charges up and tapers off at the top just like the drawing.*

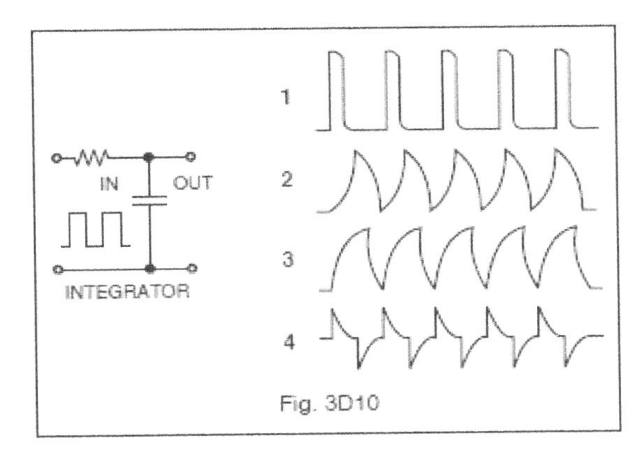

Fig. 3D10

In Figure 3D11, a pure AC signal to the input would produce an output wave on an oscilloscope that looks like: #2. *Cheat: Recognize a rectifier and the arrow points down, so the pulse is on the down side.*

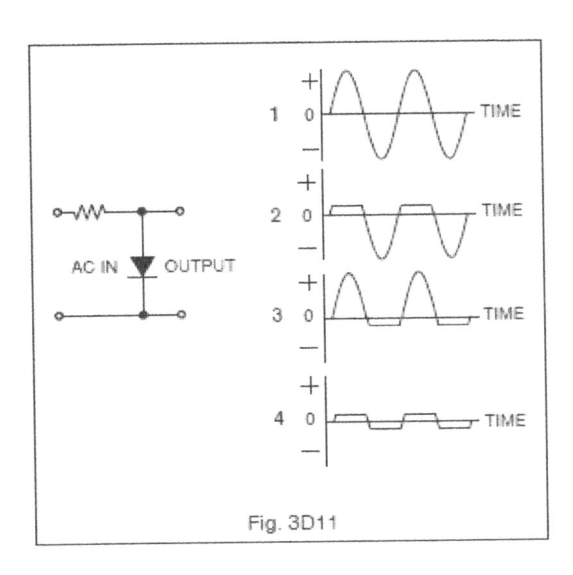

Fig. 3D11

SUMMARY Subelement D

TOPIC 29 – R-L-C CIRCUITS

Impedance at resonance = circuit resistance.
Voltage across series reactance could be higher because of resonance.
Current flow at resonance is maximum
At resonance, current and voltage are in phase.

TOPIC 30 – OP-AMPS

Op-amp better than passive filter as it provides gain.
Inverting op-amp input and output out of phase.
Gain determined by external feedback network
Op-amp in open-loop is a comparator.

TOPIC 31 – PLLS, VCOS, MIXERS

Digital to analog converter is in digital synthesizer.
Unwanted output components are spurs.
Error voltage fed back to varactor in phase-locked loop.
Phase-locked loop has a phase detector.
Mixer outputs original frequencies and sum and difference frequencies.

TOPIC 31 – SCHEMATICS

Gain of op-amp determined by resistor across it.
Diodes point way of current flow assuming positive to negative.
If coupling capacitor fails open, no amplification.
No problem with the power supply schematic.
Square wave across a capacitor will have ramp up.
Output wave across a diode will show pulse in direction of diode's arrow.

DIGITAL LOGIC (Subelement E)

33 - TYPES OF LOGIC

TTL means: transistor-transistor logic. It is binary, consisting of zeroes and ones or high or lows.

The voltage range considered to be logic low input in a TTL device operating at 5 volts is: zero to 0.8 volts. *"High" is 5 volts. "Low" is zero.*

The voltage range to be a valid high input in a TTL device operating at 5 volts would be: 2.0 to 5.5 volts.

The common power supply voltage for TTL series integrated circuits is: 5 volts. *Integrated circuits typically run on 5 volts.*

TTL inputs left open develop: a high-logic state. If we assumed "low," the slightest little jiggle in voltage might give a false "high." If we assume high, a much larger change would be required to give a false "low."

The best instrument to check a TTL logic circuit is: a DMM (digital multimeter). *The DMM gives a more precise digital reading than a VOM (Volt/Ohm meter) with a needle swinging against a scale.*

34 - LOGIC GATES

The characteristic of an AND gate is: it produces a logic "1" at its output only if all inputs are logic "1." *AND means all inputs and the output are the same.*

The characteristic of a NAND gate is: it produces a logic "0" on its output only when all inputs are logic "1." *It is Nand, so the output will be the* *reverse of the input. The AND means both inputs must be the same.*

The characteristic of an OR gate is: it produces a logic "1" at its output if any input is logic "1." *Either (or) input gives the same output.*

The characteristic of a NOR gate is: it produces a logic "0" at its output if any or all inputs are logic "1." *It is Nor so the output is reversed. The OR means either input can be a "1."*

The characteristic of a NOT gate is: it produces a logic "0" at its output when the input is logic "1" and vice versa. *NOT produces the opposite from the input.*

The logic gate which provides an active high out when both inputs are high is: an AND gate. *AND means all inputs and the output are the same*

35 - LOGIC LEVELS

In a negative logic circuit, the level used to represent a logic 0 is: high level. *It is a negative logic circuit so the normally low "0" becomes high.*

Assuming positive logic, for the logic levels shown in Figure 3E12, the logic levels of test points A, B, and C are: A is low, B is high, and C is high.

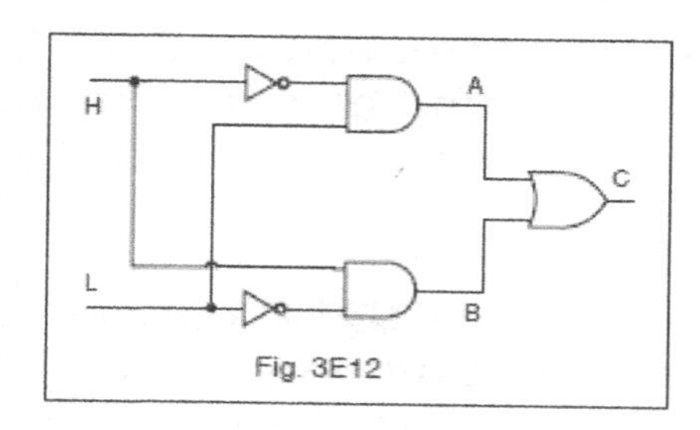

Fig. 3E12

To answer this, you must know the triangle is an inverter, the rounded object is a AND gate and the curved object is an OR gate. You could follow this through or *Cheat: Go for the Hi-C fruit punch. Look for the answer where C is high.*

Assuming positive logic, for the logic input levels in Figure 3E13, the logic levels at test points A, B, and C are: A is high, B is high, and C is high. *Solve: Those little circles on the tip of the rounded gates make them NAND gates which invert the expected answer. Cheat: All the devices are the same, so all the points are the same Hi-C.*

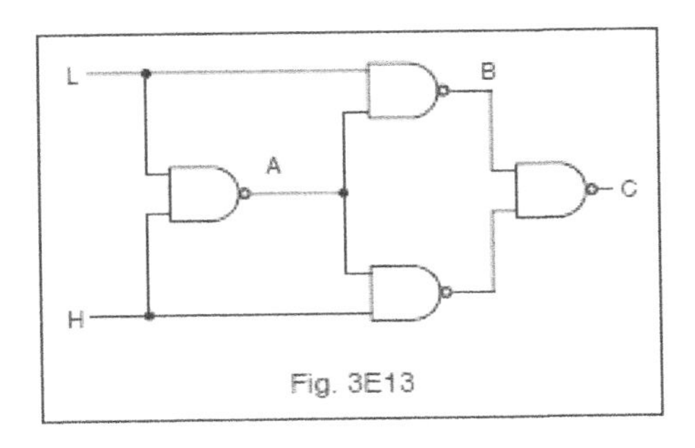

Fig. 3E13

In a positive-logic circuit, the level used to represent logic 1 is: high level.

A truth table is: a list of input combinations and their corresponding outputs that characterizes a digital devices function.

2-input AND gate

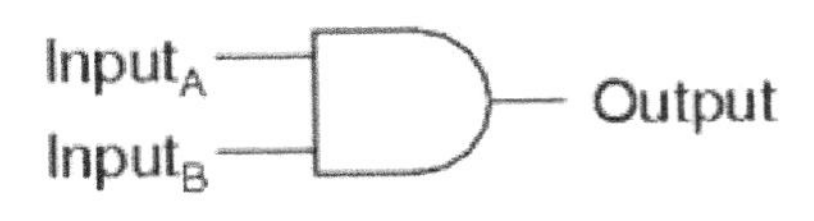

A	B	Output
0	0	0
0	1	0
1	0	0
1	1	1

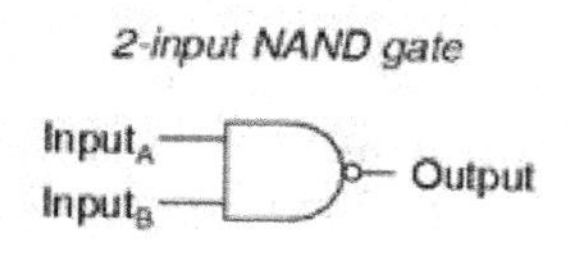

A	B	Output
0	0	1
0	1	1
1	0	1
1	1	0

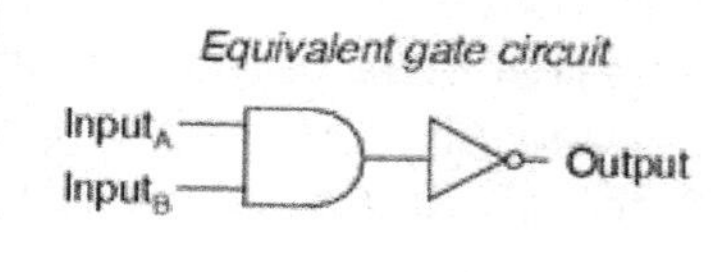

Given the input levels shown in Figure 3E14, and assuming positive logic, the output would be: A is low, B is high, and C is high. *Solve: The top gate (1) is an XOR gate, if both inputs are high, the output is low. Both inputs are high because the open line is assumed high, so the output at A is low. The bottom NAND gate (2) has an inverter in the high line so the inputs are two lows, converted to a high by the NAND gate. A high and a low into the final NAND gate (3) output a high. Cheat: Hi-C and a high B.*

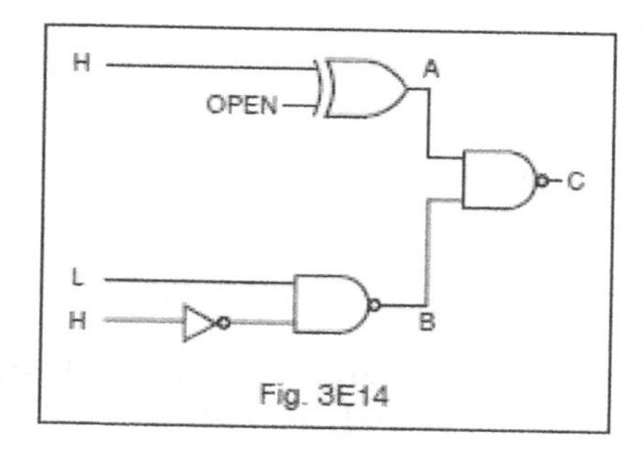

Fig. 3E14

36 - FLIP-FLOPS

A flip-flop circuit is a binary logic element with: 2 stable states. *It is binary, and that means 2.*

A flip-flop is a binary sequential logic element with: 2 states. *Same question.*

To divide a signal frequency by 4 requires: 2 flip-flops. *Each flip-flop has 2 states and 4 / 2 = 2.*

A flip-flop stores: 1 bit of information. *That bit is 1 or 0.*

To construct an 8 bit storage register would require: 8 R-S flip-flops. *1 bit in each flip-flop.* The R-S refers to "reset-set" and is not relevant to the question.

An R-S flip-flop can do everything except: operate in toggle mode with R-S inputs held constant and CLK initiated. *Too many words. R-S flip-flops don't operate in toggle mode.*

37 - MULTIVIBRATORS

The frequency of an AC signal can be divided electronically by: a bistable multivibrator. A flip-flop is a bistable multivibrator; it is stable in one of two states.

A bistable multivibrator circuit is commonly named: a flip-flop.

A bistable multivibrator circuit is: a flip-flop.

An astable multivibrator is: a circuit that alternates between two unstable states. It alternates on and off.

A monostable multivibrator is: a circuit that can be switched momentarily to the opposite binary status and then returns after a set time to the original state. It is stable in one state.

The wave form that would appear on the voltage outputs at the collectors of an astable, multivibrator common-emitter stage is: a square wave. *Too much information. An astable multivibrator alternates between two states, with no in-between, as does a square wave.*

38 - MEMORY

The name of the semiconductor memory IC[12] whose digital data can be written or read, and whose memory word address can be accessed randomly is: RAM, Random-Access Memory. *The give-away is "accessed randomly."*

The name of the semiconductor IC that has a fixed pattern of digital data stored in its memory matrix is ROM: Read Only Memory. *The giveaway is that the data is in a "fixed pattern." It can't be changed. It is Read Only.*

The term "IO" within a microprocessor system means: input-output.

The name for a microprocessor's sequence of commands and instructions is: "program." *You program the commands and instructions.*

A memory IC that has 4 data bus input/output pins and 4 address pins for connection to the address bus could contain: 64 memory cells. *Solve: The data bus is 4 bit, and each bit can be 1 or 0. The total number of combinations is 4^2 = 16. If there are 4 address pins, to access the data bus, the total memory cells are 4 x 16 = 64.*

The name of the random-access semiconductor memory IC that must be refreshed periodically to maintain reliable data storage is: DRAM, dynamic random-access memory. *It is dynamic or changing and must be refreshed.*

---

[12] IC is Integrated Circuit, a set of circuits on one chip.

39 - MICROPROCESSORS

In a microprocessor-controlled two-way radio, a "watchdog" timer: verifies the microprocessor is executing the program. *The watchdog keeps an eye on the microprocessor.*

The term "DAC" in a microprocessor circuit refers to: digital to analog converter. The digital signal must be converted to analog for you to hear and understand. Note the "D" is first, so it is digital to analog.

An MCU is a microprocessor control unit. **A voltage regulator is not part of a MCU processor.** *A voltage regulator has nothing to do with microprocessor controls.*

The portion of a microprocessor circuit that is the pulse generator is: the clock. *The clock times the pulses.*

The term "ALU" in a microprocessor means: arithmetical logic unit. *A microprocessor performs logic functions.*

The circuit that connects the microprocessor with the memory and input/output system is the data bus line. *The data travels on a bus line.*

40 - COUNTERS, DIVIDERS, CONVERTERS

The purpose of a prescaler circuit is: it divides an HF signal so that a low-frequency counter can display the operating frequency. A prescaler is an electronic counting circuit used to reduce high frequency signal to a lower frequency.

The term "BCD" means: binary coded decimal. Translating numbers to binary.

ELEMENT 3 – DIGITAL LOGIC

The function of a decade counter digital IC is: to produce one output pulse for every ten input pulses. *If it counts decades, it is counting by tens.*

The integrated circuit that converts an analog signal to a digital signal is: ADC. Analog Digital Conversion. Note the "A" is first, so this is analog to digital.

The integrated device that converts digital to analog is: a DAC. Digital Analog Conversion. Note the "D" is first. This question was asked before.

In binary numbers, the quantity two is: 0010. *Binary is all ones and zeroes. This is the only answer with 1s and 0s.*

SUMMARY – Subelement E

TOPIC 33 – TYPES OF LOGIC

TTL is transistor-transistor logic.
Voltage range for "low" is zero to .8 volts.
Voltage range for "high" is 2 – 5.5 volts.
Common power supply is 5 volts.
Open inputs treated as "high."
Check with a DMM, digital multimeter.

TOPIC 34 – LOGIC GATES

AND – outputs 1 if all inputs are 1.
NAND – outputs 0 if all inputs are 1.
OR – outputs 1 if any input is 1.
NOR – outputs 0 if any or all input are 1.
NOT – produces 0 if input is 1 and vice versa.

TOPIC 35 – LOGIC LEVELS

1 is high and 0 is low.
Truth table lists input and output combinations

TOPIC 36 – FLIP-FLOPS

Flip-Flop is a binary logic element with 2 stable states
To divide a frequency by 4 requires 2 flip-flops.
Flip-flop stores one bit of information.
8 bit storage register would require 8 flip-flops.
R-S flip-flop doesn't operate in toggle mode.

TOPIC 37 – MULTIVIBRATORS

Frequency of AC signal can be divided by bistable multivibrator (a flip-flop)
Astable vibrator alternates between two unstable states.
Monostable vibrator switches and returns.
Wave form of astable vibrator is square wave.

TOPIC 38 – MEMORY

RAM is accessed randomly.
ROM is fixed pattern of digital data.
IO means "input-output."
Sequence of commands and instructions is "program."
4 in/out pins and 4 address pins = 64 memory cells.
DRAM must be refreshed periodically.

TOPIC 39 – MICROPROCESSORS

"Watchdog" is a timer verifies microprocessor is executing the program.
DAC is Digital to Analog Converter
Voltage regulator is not part of microprocessor control.
Pulse generator is the clock.
ALU is Arithmetical Logic Unit
Circuit connecting microprocessor and memory is data-bus line.

TOPIC 40 – COUNTERS, DIVIDERS CONVERTERS

Prescaler divides HF signal for low-frequency counter
BCD is binary coded decimal.
Decade counter counts by tens.
ADC – analog to digital converter. DAC – digital to analog.
Binary 2 = 0010.

RECEIVERS (Subelement F)

41 - RECEIVER THEORY

The limiting condition for sensitivity in a communications receiver is: the noise floor of the receiver. No matter how sensitive your receiver, you can't hear much below the noise level. The noise floor discussed here is generated by the receiver.

The definition of the term "receiver desensitizing" is: a reduction in receiver sensitivity because of a strong signal on a nearby frequency. A strong signal close to yours can overwhelm the receiver even if you can't hear it.

The term used to refer to a reduction in receiver sensitivity caused by unwanted high-level adjacent channel signals is: desensitizing. *Same question.*

The term noise figure of a communications receiver is: the level of noise generated in the front end and succeeding stages of a receiver. *The noise figure of the receiver is a measure of the noise generated in the receiver.*

The stage of a receiver that primarily establishes its noise figure is: the RF stage. The RF stage is the most sensitive and the most sensitive to noise.

The term for the ratio between the largest tolerable receiver input signal and the minimum discernible signal is: the dynamic range. *Dynamic range is a ratio of two signal levels, large and small.*

42 - RF AMPLIFIERS

Selectivity can be achieved in the front-end circuitry of a communications receiver by using: a preselector. *It pre-selects the signals to provide additional selectivity.*

The primary purpose of an RF amplifier in a receiver is: to improve the receiver's noise figure. The RF amplifier is designed to improve the signal without introducing noise from within the receiver circuits. It is not to provide most of the receiver gain.

The amount of gain to use in the RF amplifier stage of a receiver is: sufficient gain to allow weak signals to overcome noise generated by the first mixer stage. *Sufficient signal gain to be heard over the receiver noise.*

Too much gain in a VHF receiver could result in: susceptibility of intermodulation interference from nearby transmitters. *Too much gain overloads the receiver.*

The advantage of a GaAsFET preamplifier in a modern VHF radio receiver is: high gain and low noise floor. Gallium Arsenide Field Effect Transistors have low noise and high gain.

In a VHF receiver, a low noise amplifier would be most advantageous: in the front end RF stage. *Boost the signal at the input but add low noise.*

43 - OSCILLATORS

A Colpitts oscillator is commonly used in a VFO (variable frequency oscillator) because: it is stable. *Stability is very important.*

The oscillator stage in a frequency synthesizer is called: a VCO. *VCO is voltage controlled oscillator. The oscillator stage has to include the word oscillator.*

The three major oscillator circuits found in radio equipment are: Colpitts, Hartley, and Pierce. *President Taft has nothing to do with it. All the other answers include Taft.*

The oscillator commonly used in a VFO is a: Colpitts. A Colpitts is controlled by a variable capacitor, and they were easy to come by.

For a circuit to oscillate: it must have sufficient positive feedback. Feedback is what gets an oscillator going. Negative feedback would cancel out the oscillations.

In Figure 3F15, the symbol representing a local oscillator is: 2. Circle 1 is a mixer, and the local oscillator is injecting a signal to produce the IF, intermediate frequency signal.

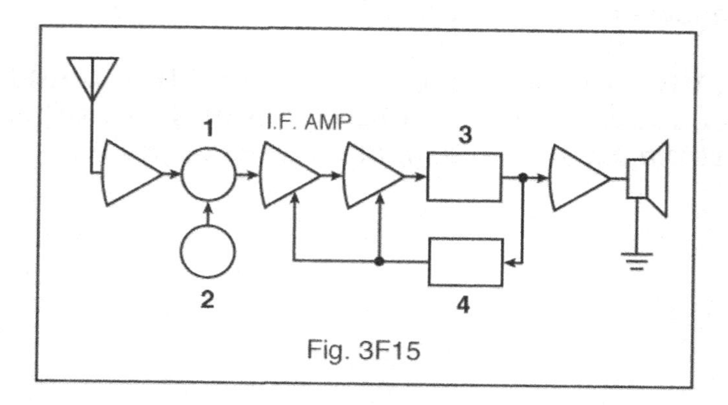

Fig. 3F15

44 - MIXERS

The mixing process in a radio receiver is: the combination of two signals to produce sum and difference frequencies.

The principal frequencies that appear at the output of a mixer circuit are: the original frequencies and the sum and difference frequencies. This is a repeat question.

If the normal channel is 151.000 MHz, the IF is operating at 11.000 MHz, and the LO[13] is at 140.000 MHz, the image frequency is: 129.000 MHz. *Mixing produces sums and differences. The image is the difference in mixing of 140 and 11 or 129.*

The radio stage where the image frequency is normally rejected is the: RF stage. *The mixing is in the RF stage before it goes to the IF (intermediate frequency) stage. The rejecting needs to be done in the RF stage.*

If a receiver mixes a 13.8 MHz VFO with a 14.255 MHz receive signal to produce a 455 kHz intermediate frequency, a 13.345 MHz signal will produce: an image response. 13.8 - 455 = 13.345. *The signal is an image. It is not real.*

If excessive amounts of signal overdrive the mixer circuit: spurious mixer products are generated. *Overdriving introduces distortion also known as spurious products.*

---

[13] LO refers to local oscillator.

45 - IF AMPLIFIERS

The degree of selectivity desirable in the IF circuitry of a wideband FM receiver is: 15 kHz. *Wideband FM is wide. Pick the widest choice.*

The filter to use in a micro-miniature electronic circuit is: a receiver SAW IF filter. SAW is Surface-Acoustic-Wave, and the part is very small. *Cheat: If you see saw, you have your answer.*

A receiver selectivity of 2.4 kHz in the IF circuitry is: optimum for SSB voice. That is the approximate width of a single sideband signal.

A receiver selectivity of 10 kHz in the IF circuitry optimum for: double sideband AM. Double sideband AM is more than twice as wide as single sideband. The appropriate trigger word is AM.

The undesirable effect of using too wide a filter bandwidth in the IF section of a receiver is: undesired signals will reach the audio stage. *Too wide of a filter lets in interference.*

The filter bandwidth of the receiver should be: slightly greater than the received-signal bandwidth. Too little, and the signal sounds pinched. Too wide, and you get interference.

46 - FILTERS AND IF AMPLIFIERS

The primary purpose of the final IF amplifier stage in a receiver is: gain. *An amplifier produces gain.*

When selecting an intermediate frequency, consider: image rejection and selectivity. *Image rejection and selectivity are key.*

The primary purpose of the first IF amplifier stage in a receiver is: selectivity. *Knock down the interference before it gets into later stages of the radio.*

The parameter to select when designing an audio filter using an op-amp is: bandpass characteristics. *If you are designing a filter, you are concerned with bandpass characteristics.*

The distinguishing feature of a Chebyshev filter is: it allows ripple in the passband. Not good.

It would be desirable to use an m-derived filter over a constant-k filter when: you need more attenuation at a certain frequency that is too close to the cut-off for a constant-k filter.

M–derived filters have a sharp cut-off but lose effectiveness as you get away from the design frequency. Constant-k filters lack the sharp-cutoff but they are effective over a wider range. Constant-ks are used to suppress harmonics which would be far from the design frequency.

47 - FILTERS

A good crystal band-pass filter for single-sideband phone would be: 2.1 kHz. This is the closest answer to the 2.4 kHz suggested on the previous page.

The three general groupings of filters are: high-pass, low-pass and band-pass. They pass everything above the cutoff frequency (high-pass), below the cut-off frequency (low-pass) or between two frequencies (band-pass).

There is another group called "band-stop" that notch out between two frequencies. The next question has a band stop filter.

Regarding Figure 3F16: A is a low pass curve and D is a band stop curve. *A has a strong signal below the cut-off. D has a notch which is the band being stopped.*

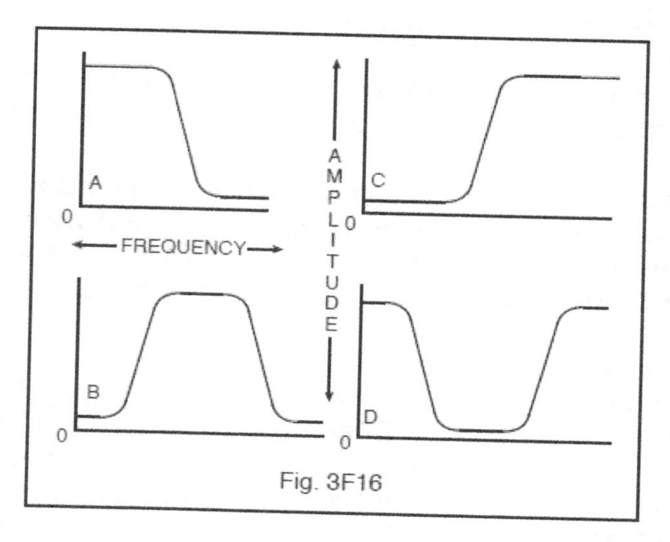

Fig. 3F16

An m-derived filter: uses a trap to attenuate undesired frequencies too near the cutoff for a constant-k filter. *Variation on a question from the last section. M-derived filters have a sharp cut-off.*

The advantage of a constant-k filter is: it has high attenuation of signals at frequencies far removed from the pass band. Constant-k filters are used to reject harmonics.

The distinguishing feature of a Butterworth filter is it has a maximally flat response over its passband. No ripples like a Chebyshev. *Cheat: Mrs. Butterworth's syrup is smooth.*

48 - DETECTORS

The definition of detection in a radio receiver is: the recovery of intelligence from the modulated RF signal. *You are detecting the meaning of the signal.*

A product detector: uses a mixing process with a locally generated carrier. *You need a locally generated carrier in a product detector to add back the carrier and make single sideband understandable.*

The circuit used to detect FM-phone signals is: a frequency discriminator. *FM is frequency modulation, and demodulating is done by a frequency discriminator.*

A frequency discriminator in a radio receiver is: a circuit for detecting FM signals.

The process of detection in a radio diode detector circuit is: rectification and filtering of RF. *A diode is a rectifier.*

In a CTCSS controlled FM receiver: the CTCSS tone is filtered out after the discriminator but before the audio section. Continuous Tone Coded Squelch System sends a sub-audible tone. You must have that tone programmed in your receiver to hear the transmission. The tone is filtered out before the audio section.

49 - AUDIO & SQUELCH CIRCUITS

The digital processing term for noise subtraction circuitry is: adaptive filtering and autocorrelation. *Digital signal processing uses adaptive filtering meaning the processing adapts around the signal to shut out noise and interference.*

The purpose of de-emphasis in the receiver audio stage is when coupled with transmitter pre-emphasis, flat audio and noise reduction is received. *You want flat audio and noise reduction.* De-emphasis and pre-emphasis compliment each other.

Digital Coded Squelch works: because of digital codes. *It is digitally coded, so it relies on digital codes.*

A squelch circuit functions because of: the presence of noise. *Noise mutes the receiver. A signal opens the receiver.*

A CTCSS squelch works: based on tones. *Continuous Tone-Coded Squelch System uses tones.*

The radio circuit that samples analog signals, records and processes them as numbers, then converts them back to analog signals is: a digital signal processing circuit. *The circuit processes digital and analog.*

50 - RECEIVER PERFORMANCE

You would normally find a low-pass filter in: a radio receiver's AVC circuit and Power Supply. *Not the oscillator because it is already tuned. Cheat: Another question where the choice of two of the answers is correct.*

To use ferrite beads to suppress ignition noise, install them: in the primary and secondary ignition leads. *To suppress ignition noise, install on the ignition leads.*

The term used to refer to the condition where signals from a very strong station are superimposed on other signals is: cross-modulation interference. *The two signals are mixing and cross-modulating.*

Cross-modulation interference is: modulation from an unwanted signal heard in addition to the desired signal. *Cross-modulation is modulation.*

In Figure 3F15, the point in the circuit a DC voltmeter could be used to monitor signal strength is: point 4. This is the AGC or automatic gain control. You can see by the arrows it feeds back to adjust the gain in the IF amplifier to keep the signals at a constant level. The feedback voltage is a measure of signal strength.

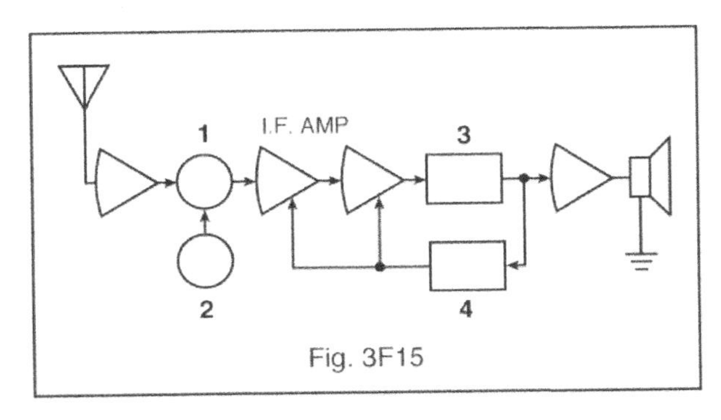

Fig. 3F15

Pulse type interference to automobile radio receivers that appears related to the speed of the engine can be reduced by: installing resistances in series with the spark plug wires.
Pulses related to speed must be from the spark plugs.

SUMMARY Subelement F

TOPIC 41 – RECEIVER THEORY

Sensitivity is limited by noise floor of the receiver.
Desensitizing is caused by a nearby strong signal.
Noise figure is level of noise generated in receiver.
RF stage establishes noise figure.
Dynamic range is ratio between strongest and weakest discernable signal.

TOPIC 42 – RF AMPLIFIERS

Pre-selector improves selectivity in front-end.
Purpose of RF amplifier is to improve noise figure.
Use sufficient gain so weak signals overcome noise. generated in the first mixer stage.
Too much gain can cause intermodulation interference.
GaAsFET – high gain, low noise floor.
Low noise amplifier best in front end RF stage.

TOPIC 43 – OSCILLATORS

Colpitts oscillator used in a VFO.
VCO (Voltage Controlled Oscillator) used in a frequency synthesizer.
Oscillators are Colpitts, Hartley and Pierce.
Must have sufficient positive feedback to oscillate.
Local oscillator is at the first mixer.

TOPIC 44 – MIXERS

Mixing is combining two frequencies to produce sum and difference.
Output of mixer is original frequencies plus sum and difference.
Image frequency is unintended mixes.

RF Stage is where image is normally rejected.
Excessive overdrive produces spurious mixer products.

TOPIC 45 – IF AMPLIFIERS

Selectivity for wideband FM is 15 kHz.
SAW-IF filter is micro-miniature.
Selectivity for SSB is 2.4 kHz.
Selectivity for double sideband AM is 10 kHz.
Too wide a filter allows undesired signals.
Filter bandwidth should be slightly greater than signal.

TOPIC 46 – FILTERS AND IF AMPLIFIERS.

Final IF amplifier is for gain.
First IF amplifier is for selectivity.
To choose an IF consider image rejection and selectivity.
Design a filter considering bandpass characteristics.
Chebyshev filter has ripple.
M-derived filter has steeper skirts than constant-k.

TOPIC 47 – FILTERS

Good crystal filter for SSB is 2.1 kHz.
3 types of filters: high-pass, low-pass and band-pass.
Constant-k filter works over wide range.
Butterworth filter is smooth.

TOPIC 48 – DETECTORS

Detection is recovery of intelligence.
Product detector uses mixing process.
Detect FM phone with a frequency discriminator.
Radio diode detection circuit uses rectification and filtering.
CTCSS (Continuous Tone Coded Squelch System) tone if filtered out before the audio circuit.

TOPIC 49 – AUDIO & SQUELCH CIRCUITS

Digital processing uses adaptive filtering and autocorrelation.
De-emphisis and pre-emphasis used for flat audio and noise reduction.

ELEMENT 3 - RECEIVERS

Digital Coded Squelch relies on digital codes.
Squelch circuit works because of presence of noise.
CTCSS squelch works based on tones.
Analog processed as numbers then back to analog is digital signal processing.

TOPIC 50 – RECEIVER PERFORMANCE

Low pass filter in AVC circuit and power supply.
To suppress ignition noise, install ferrite beads on ignition leads.
Cross-modulation is strong signals superimposed on other signals. Unwanted signal heard in addition to the desired signal.
Monitor signal strength at the AGC feedback circuit.
Reduce pulse interference related to speed of auto's engine by installing resistance wires on spark plugs.

TRANSMITTERS (Subelement G)

51 – AMPLIFIERS – 1

We classify amplifiers according to how much of the 360-degree signal cycle they operate. For example, Class As amplify throughout the entire cycle. Class Cs only through a part of the cycle.

The class of amplifier characterized by the presence of output throughout the entire signal cycle and the input never goes into the cutoff region is: Class A. *Too many words. If the amplifier is "on" through the entire cycle, it is Class A.*

The distinguishing feature of a Class A amplifier is: output for the entire 360 degrees of the signal cycle.

The type of amplifier with the highest linearity and least distortion is: Class A. *It is "on" through the entire cycle and leaves nothing out.*

The class of amplifier with the highest efficiency is: Class C. *Cheat: "Efficen-cee."* Class C is not "on" for most of the cycle, and the amplifier rests, so it is more efficient.

The class of amplifier distinguished by the bias being set well beyond cutoff is: Class C. The bias is set such that the amplifier only works for part of the cycle.

The class of amplifier with an operating angle of more than 180 degrees but less than 360 degrees is: AB. *More than 180 so it can't be C. Less than 360 so it is not A. B is exactly 180. AB is the remaining answer.*

52 – AMPLIFIERS – 2

The Class B amplifier output is present: for 180 degrees. Class B amplifiers operate exactly half the cycle and are placed back-to-back in a push-pull arrangement. The output is similar to Class A, but each side only works half the time. *Cheat: Back-to-back is Class B*

The input-amplitude parameter most valuable in evaluating the signal handling capability of a Class A amplifier is: peak voltage. *If you are evaluating input amplitude capability, you are concerned about the most drive the amplifier can take, peak voltage.*

In a Class C amplifier: output is present for less than 180 degrees of the input cycle.

The approximate DC power input to a Class AB RF power amplifier stage in an unmodulated carrier transmission when the PEP output power is 500 watts is: 1000 watts. *Way too much complication. Class AB is about 50% efficient, so the DC power input is twice the output.*

The Class AB amplifier output is: present for more than 180 but less than 360 degrees. *Same question as in the previous Topic.*

The class of amplifier characterized by conduction for 180 degrees of the input wave is: Class B.

Class	Degrees	Efficiency
A	360	<50%
B	180	60%
C	<180	80%
AB	>180, <360	50%

53 - OSCILLATORS AND MODULATORS

The frequency deviation of an FM phone signal is greater than the modulating frequency. The ratio is the modulation index.

The modulation index in an FM phone signal having a maximum frequency deviation of 3,000 Hz on either side of the carrier frequency when the modulating frequency is 1,000 Hz is: 3. *Solve: To find the modulation index divide the deviation by the modulation: 3000/1000 = 3.*

The modulation index of an FM phone transmitter producing a maximum carrier deviation of 6 kHz when modulated with a 2 kHz modulating frequency is: 3. *Solve: 6/2 = 3.*

The total bandwidth of an FM phone transmission having a 5 kHz deviation and a 3 kHz modulating frequency is: 16 kHz. *Solve: Bandwidth is 2 times the total of the deviation and modulation. 2 X (5+3) = 16. It is 2 times because of the plus and minus sides of the FM signal.*

The modulation index of a phase-modulated emission: does not depend on the RF carrier frequency. Phase-modulation is FM and the modulation index, defined above, does not take into account carrier frequency.

A single-sideband phone signal can be generated by: using a balanced modulator followed by a filter. The balanced modulator nixes the carrier, and the filter takes out the unwanted sideband. *Filter out the unwanted sideband.*

A balanced modulator: produces a double sideband, suppressed carrier signal. The balanced modulator only nixes the carrier. You still need a filter to get single sideband.

54 - RESONANCE – TUNING NETWORKS

An L-network is: a network consisting of an inductor and a capacitor. One is in series, and one is in parallel to the input. The shape is an "L."

A pi-network is: one inductor and two capacitors or one capacitor and two inductors. *The connection looks like the Greek letter Π.*

The resonant frequency of an electrical circuit is: the frequency at which capacitive reactance equals inductive reactance. The two cancel each other.

The three network types commonly used to match an amplifying device to a transmission line are: L network, pi network and pi-L network.

A pi-L network is: a network consisting of two capacitors and two inductors. *Two components form an L. Three components form Π. Add a component to the Pi and get a Pi-L.*

The network that provides the greatest harmonic suppression is: a Pi-L. *More components equals more suppression.*

55 - SSB TRANSMITTERS

If a non-linear amplifier is used with a single-sideband phone transmitter: you will get distortion. *If the amplifier is not linear, it will distort.*

To produce a single-sideband suppressed carrier transmission: it is necessary to cancel the carrier and filter the unwanted sideband.

The PEP-to-average power ratio in a single-sideband signal is determined by: the speech

characteristics. *A monotone will make the PEP (Peak Envelope Power) and average almost the same.*

The approximate ratio of peak envelope power to average power during normal voice modulation in a single-sideband signal is: about 2.5 to 1. *Peak is about 2.5 times average. The other answers are way off.*

The output peak envelope power from a transmitter as measured on an oscilloscope showing 200 volts peak-peak across a 50-ohm resistor is: 100 watts. *Solve: This is AC. To use Ohm's law, we must convert to DC. First, convert peak-to-peak to peak by dividing by 2. E = 100. Multiply by .707 to get the RMS voltage = 70.7.* *$P = E^2/R$. P = 70.7 x 70.7 = 4998 / 50 = 100.*

The voltage across a 50-ohm dummy load dissipating 1,200 watts would be: 245 volts. *Solve:* $E^2 = PR$. *Voltage is the square root of power times resistance. 50 x 1200 = 60,000. The square root of 60,000 is 245.*

56 - TECHNOLOGY

Intermodulation interference between two transmitters in close proximity can be reduced or eliminated by: installing a terminated circulator or ferrite isolator in the feed line to the transmitter and duplexer. *To isolate the transmitters, install an isolator.*

Parasitic oscillations can be eliminated in a power amplifier by: neutralization. *Neutralize the parasites.* Neutralization is introducing negative feedback to cancel the parasitic oscillations.

The name of the condition when signals of two transmitters in close proximity mix together in one or both of their final amplifiers, and

unwanted signals at the sum and difference frequencies of the original transmissions are generated is: intermodulation interference. *Unwanted mixing of signals is intermodulation.*

A wide-band communications system in which the RF carrier varies according to some pre-determined sequence is: spread-spectrum communications. *The RF carrier frequency varies, so the signal is spread out. The "pre-determined sequence," allows the receiver to follow.*

The modulation type what can be frequency hopping of one carrier or multiple simultaneous carriers is: spread spectrum. *Frequency hopping is spread spectrum.*

Even-order harmonics can be reduced or eliminated in transmitter amplifier design by: using a push-pull amplifier. Class B, push-pull, reduces even harmonics.

SUMMARY Subelement G

TOPICS 51 and 52 – AMPLIFIERS

Class A entire cycle, highest linearity, signal-handling ability defined by peak voltage.
Class B 180 degrees.
Class AB between 180 and 360 degrees, 50% efficient.
Class C less than 180 degrees, highest efficiency, bias beyond cutoff.

TOPIC 53 – OSCILLATORS AND MODULATORS

FM Modulation Index = deviation / modulation, does not depend on carrier frequency.
FM total bandwidth = 2 times sum of deviation and modulation.
SSB generated by balanced modulator and filter.
Balanced modulator produces double sideband, suppressed carrier.

TOPIC 54 – TUNING NETWORKS

L-network is inductor and capacitor
Pi-network has 3 components.
Pi-L has four components, greatest harmonic suppression.
Resonant frequency is where inductive and capacitive reactance are equal.

TOPIC 55 – SSB TRANSMITTERS

Non-linear amplifier will produce distortion.
To get SSb, cancel carrier and filter one sideband.
PEP/average power ratio determined by speech characteristics. Normally 2.5:1.
Convert AC to RMS and then apply Ohm's Law. Divide peak-to-peak by 2 to get peak, multiply by .707.
$P = E^2/R$ E = square root of PR

TOPIC 56 – TECHNOLOGY

Reduce intermod with terminated circulator or ferrite isolator in the feedline.
Parasitic oscillations eliminated by neutralization.
Two signals mixing and generating unwanted signals is intermodulation.
Wide-band system where carrier varies in predetermined sequence is spread spectrum.
Frequency hopping is spread spectrum.
Class B, push-pull amplifier reduces even harmonics.

MODULATION (Subelement H)

57 – FREQUENCY MODULATION

The deviation ratio is: the maximum carrier frequency deviation to the highest audio modulating frequency. *The deviation ratio compares deviation to something. Look for deviation in the answer. The deviation ratio is the ratio of the maximum deviation to the maximum modulating frequency.*

The deviation ratio for an FM phone signal having a maximum frequency deviation of plus or minus 5 kHz and accepting a maximum modulation rate of 3 kHz is: 1.66. *Solve: Divide the maximum deviation by the modulation rate. 5 / 3 = 1.66.*

The deviation ratio of an FM-phone signal having a maximum frequency swing of plus or minus 7.5 kHz accepting a maximum modulation of 3 kHz is: 2.14. *Solve: 7.5 / 3.5 = 2.14.*

An FM-phone signal can be produced in a transmitter by: feeding the audio directly to the oscillator. *Feeding audio to the oscillator would change the frequency output (frequency modulation).*

The modulation index is: the ratio between the deviation of a frequency modulated signal and the modulating frequency. The modulation index is similar to the deviation ratio, but the deviation ratio is measured against the maximums.

The term for the maximum deviation from the carrier frequency divided by the maximum audio modulating frequency is: deviation index. *Deviation index (ratio) is based on maximums.*

58 - SSB MODULATION

In Figure 3H17, the block labeled 4 would indicate that this schematic is most likely: an SSB radio transmitter. *Cheat: Ignore block 4. Recognize this as an SSB transmitter because of the microphone at 2 and antenna on the output.*

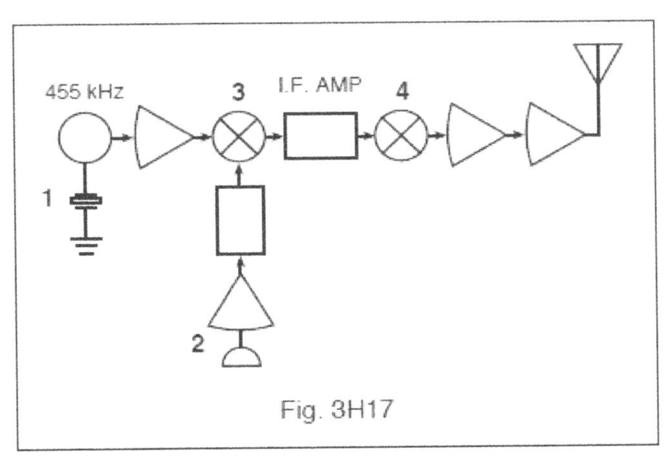

Fig. 3H17

The block where the audio intelligence is inserted is: #2. A microphone.

A two-tone test feeds two non-harmonically related signals into the SSB transmitter and observes the output pattern on an oscilloscope. The pattern should be symmetrical. **To test the linearity of a single-sideband phone transmitter while viewing the output on an oscilloscope: use a two-tone audio sine wave.**

A two-tone test illustrates on an oscilloscope: the linearity of an SSB transceiver.

The signals used to conduct an SSB two-tone test: are two non-harmonically related audio signals.

Double sideband signals can be produced by modulating the supply voltage to a Class C amplifier. *Audio signals produced by modulating.*

59 – PULSE MODULATION

An important factor in pulse-modulation using time-division multiplex is: synchronizing of transmit and receive clock pulses. *If it is time division, you need a clock.*

In a pulse-width modulation system, the parameter the modulating signal varies is: the pulse duration. *Pulse width is another way of saying pulse duration.*

The name of the type of modulation in which the modulating signal varies the duration of the transmitted pulse is: pulse-width modulation.

A pulse modulation system is best described: as the peak transmitter power is normally much greater than the average power. *The transmitter is sending pulses and is silent between them. The average is less than the peak.*

In a pulse-position modulating system, the parameter the modulating signal varies: is the time at which each pulse occurs. *Pulse position involves timing, not duration or number. The question is about pulse position, not pulse width.*

One way voice is transmitted in a pulse-width modulation system is: a standard pulse is varied in duration by an amount depending on the voice waveform at that instant. *Pulse width means the pulse is varied in duration. None of the other answers mentions duration.*

SUMMARY – Subelement H

TOPIC 57 – FREQUENCY MODULATION

Deviation ratio is maximum carrier deviation/highest audio frequency.
Can produce FM by feeding audio to oscillator.
Modulation index is ratio of deviation and modulating frequency. (Not maximums)

TOPIC 58 – SSB MODULATION

Identity schematic of SSB transmitter by microphone and antenna.
Audio intelligence inserted by microphone.
Test linearity with two-tone sine wave.
Two-tone test is on an oscilloscope.
Can produce double sideband by modulating power supply of Class C amplifier.

TOPIC 59 – PULSE MODULATION

Synchronize transmit and receive clock pulses.
Pulse-width modulation varies pulse duration.
Peak power much greater than average.
Pulse-position modulation varies timing.
Pulse-width varies depending on voice waveform.

POWER SOURCES (Subelement I)

60 – BATTERIES – 1

When a lead-acid battery is being charged, a harmful effect to humans is: emission of hydrogen gas.

A battery with a terminal voltage of 12.5 volts is to be trickle charged at a 0.5 A rate. The resistance to be connected in series with the battery to charge it from a 110-V DC line: is 195 ohms. *Solve: The voltage drop is 110 – 12.5 = 97.5. Applying Ohm's law E/I = R. 97.5 / .5 = 195.*

The capacity in amperes a storage battery needs to operate a 50-watt radio transmitter for 6 hours assuming a continuous transmitter load of 70% of the key-locked demand of 40 A, and an emergency light load of 1.5 A is: 177 ampere-hours. *Solve: Ignore the transmitter output; it doesn't matter. You know you need 40 amps 70% of the time which is 28 amps, plus the light of 1.5 A equals an adjusted load of 29.5 amps. To last 6 hours, multiply 29.5 x 6 = 177 ampere-hours.*

A nickel-cadmium cell has an operating voltage of about: 1.25 volts.

The total voltage when 12 Nickel-Cadmium batteries are connected in series is: 15 volts. *Solve: When connecting batteries in series, add the voltages. Ni-cads are 1.25 volts each so 12 of them would be 15 volts.*

The average fully-charged voltage of a lead-acid storage cell is: 2.06 volts. A car battery is 6 cells connected in series for a total voltage of 12.36 volts.

61 – BATTERIES – 2

If an emergency transmitter uses 325 watts and a receiver uses 50 watts, a 12.6 volt, 55 ampere-hour battery supply full power to both units for: 1.8 hours. *Solve: The load is 325 + 50 watts = 375 watts. At 12.6 volts, the amperage would be I = P/E = 30 amps. The supply is good for 55 amp hours, so 55/30 = 1.8 hours.*

The current flow in a 6-volt storage battery with an internal resistance of .01 ohms when a 3-watt, 6-volt lamp is connected will be: 0.4995 amps. *Solve: First, the resistance of the lamp will be R = E^2/P or 36/3 = 12 ohms. Add the .01 ohm from the battery and the total resistance is 12.01 ohms. I=E/R so I = 6/12.02 = .4995.*

A 6-volt battery with 1.2 ohms internal resistance is connected across two light bulbs in parallel whose resistance is 12 ohms each; the current flow is: .83 amps. *Solve: The resistance of the two 12 ohm lights in parallel is 6 ohms. Their 6 ohms plus the 1.2 ohms in the battery gives a total resistance of 7.2 ohms. I=E/R. I = 6/7.2 = .83 amps*

A ship RADAR unit uses 315 watts, and a radio uses 50 watts. If the equipment is connected to a 50 ampere-hour battery rated at 12.6 volts the battery will last: 1 hour 43 minutes. *Solve: Total load is 315+50 = 365 watts. At 12.6 volts that is P/E = I or 365 / 12.6 = 29 amps. 50 /29 = 1.7 hours which is 1 hour and 43 minutes.*

A marine radiotelephone receiver uses 75 watts of power and a transmitter uses 325 watts, they both can operate on a 50 ampere-hour, 12-volt battery for: 1½ hours. *Solve: Total load is 400 watts. At 12 volts that is P/E=I or 400/12 = 33.3 amps. 50 / 33.3 = 1 ½ hours.*

A 12.6 volt, 8 ampere-hour battery is supplying power to a receiver that uses 50 watts and a RADAR system that uses 300 watts. The battery will last 17 minutes. *Solve: Total load is 350 watts. At 12.6 volts, that is P/E = I or 350/12.6 = 27 amps. The battery is good for 8 amps so it will deplete in 8/27 = .29 hour or 60 X .29 = about 17 minutes.*

62 - MOTORS & GENERATORS

If the load is removed from an operating series DC motor: it will accelerate until it falls apart.

If a shunt motor running with a load has its shunt field opened: the speed of the motor will speed up. *Cheat: Both answers speed up.*

The expression "voltage regulation" as it applies to a shunt-wound DC generator operating at a constant frequency refers to voltage fluctuations from load to no-load. *Voltage regulation refers to voltage fluctuations.*

The line current of a 7 horsepower motor operating on 120 volts at full load, a power factor of 0.8 and 95% efficient would be: 57.2 amps. *Solve: One horsepower is 746 watts, so the motor is 7 x 746 = 5222 watts. Divide by the power factor and the efficiency to get the actual input power consumed = 6871.05 watts input divided by the actual line voltage of 120 = 57.2 amps.*

A 3 horsepower, 100 V DC motor is 85% efficient and the current is: 26.3 amperes. *Solve: 3 horsepower is 3 x 746 = 2238 watts. Divide by the efficiency shows you put 2633 watts in. I = P/E I = 2633/100 = 26.3 amps.*

The output of a separately-excited AC generator running at a constant speed can be controlled by: the amount of field current. Field current is a kind of feedback. It can vary as needed. *The other answers are fixed components.*

SUMMARY Subelement I

TOPIC 60 AND 61 – BATTERIES

Danger of lead-acid battery is hydrogen gas.
Resistor for charge rate is voltage drop E/I
Total amps needed x hours = amp hours.
Nickel-Cadmium battery is 1.25 volts.
Twelve NiCads = 12 X 1.25 = 15 volts.
Lead-acid storage cell is 2.06 volts.
Amps = total load in watts divided by voltage I=P/E
Divide amps into battery amp-hours to get battery life.

TOPIC 62 – MOTORS & GENERATORS

If load removed from series DC motor, it accelerates until it falls apart.
If shunt motor has shunt field opened, it speeds up.
One horsepower is 746 watts. To determine amps, divide watts by power factor and efficiency to get total input watts needed. Total amps is I = P/E
Output of AC generator can be controlled by the amount of field current.

ANTENNAS (Subelement J)

63 – ANTENNA THEORY

A high standing wave ratio on a transmission line could be caused by: a detuned antenna coupler. *The antenna coupler matches the antenna to the line or the line to the transmitter. If it is detuned, high SWR could result.*

The term radiation resistance of an antenna refers to: the equivalent resistance that would dissipate the same amount of power as that radiated from an antenna. *Radiation resistance is equivalent resistance.*

The value of the radiation resistance of an antenna is important because: knowing the radiation resistance makes it possible to match impedances for maximum power transfer. *Matching impedances transfer maximum power.*

A radio frequency device that allows RF energy to pass through in one direction with very little loss but absorbs RF power in the opposite direction is: an isolator. *The device isolates the input and output.*

The advantage of a trap antenna is: it may be used for multiband operation. Traps, tuned circuits, in the antenna electrically isolate portions, so the antenna length is correct for the frequency used.

Antenna bandwidth refers to: the frequency range over which the antenna can be expected to perform well.

64 – VOLTAGE, CURRENT AND POWER RELATIONSHIPS

The current flowing through a 52-ohm line with an input of 1,872 watts is: 6 amps. *Solve* $P=I^2R$ *or* $I^2=P/R$. *1872/52= 36. The square root of 36 is 6.*

The voltage produced in a receiving antenna is: always proportional to the received field strength. *The stronger the field, the more voltage produced in the receiving antenna.*

The best standing wave ratio is 1:1.

At the ends of an antenna, the values of voltage and current are: maximum voltage and minimum current. *There can't be any current at the end of a wire. It has nowhere to go. There is maximum voltage because there is zero current flow.*

If an antenna radiates a primary signal or 500 watts output, and a 2nd harmonic output of 0.5 watts, the attenuation of the second harmonic is: 30dB. *Solve: The attenuation is 500/.05 or 1000 times. Decibels are logarithmic, power of 10. 10 times is 10 dB. 100 times is 20 dB and 1000 times is 30 dB.*

An improper impedance match between a 30-watt transmitter and the antenna, with 5 watts reflected results in: 25 watts actually radiated. *30 watts out and 5 back leaves 25 watts to radiate.*

65 - FREQUENCY AND BANDWIDTH

A vertical antenna receives signals: equally from all horizontal directions. *A vertical receives equally poorly in all directions.*

The resonant frequency of a Hertz antenna can be lowered by: placing an inductance in series with the antenna. *Adding inductance lengthens the antenna and lowers the resonant frequency.*

To lengthen an antenna electrically: add a coil.

To increase the resonant frequency of a ¼ wavelength antenna: add a capacitor in series. *Adding a capacitor shortens the electrical length and raises the resonant frequency.*

An excited ½ wavelength antenna produces: both electro-magnetic and electro-static fields. All antennas produce both electro-magnetic and electro-static fields.

As an antenna is shortened through the use of loading coils: the bandwidth decreases. There is the trade-off. Loading a shortened antenna with a coil may get you resonant, but the bandwidth decreases

66 – TRANSMISSION LINES

The velocity factor of transmission line is: the ratio of the velocity of the wave in the transmission line divided by the velocity of light in a vacuum. Transmission line slows the wave. The ratio is always less than 1:1 and the velocity of light goes on the bottom of the equation. The velocity factor of coaxial cable is typically .67

The velocity factor is determined by: the dielectrics of the line. "Dielectrics" refers to the insulation separating the center conductor and the shield.

Nitrogen is placed in transmission lines: to prevent moisture from entering. Moisture is the great enemy of transmission line.

If a perfect (no loss) coaxial cable has 7 dB of reflected power when the input is 5 watts. The output of the transmission line is: 1 watt. *Solve: 3 dB is half, 6 dB is half again or one-quarter. Less than one-quarter of the 5 watts would make it through the cable, and the answer is 1 watt.*

A shorted stub attached to the transmission line to absorb even harmonics would be: ¼ wavelength. A ¼ wavelength line would be ½ wavelength on the harmonic. The impedance at the end of a ½-wave line mirrors the impedance at the other end. The shorted line will short the harmonic, so it is not radiated.

If a transmission line has a power loss of 6 dB per 100 feet, the power at the end of a 200-foot transmission line fed by a 100-watt transmitter would be: 6 watts, *Solve. 200 feet would have twice the loss as 100 or 12 dB. 3dB is half, six dB is half again and 9 db is half that or 1/8 and 12 dB would be 1/16. The power would be 1/16 of 100 watts or 6 watts. Another way to look at it is 10 dB is one-tenth, so the answer has to be less than 10.*

67- EFFECTIVE RADIATED POWER

The effective radiated power of a repeater with 50 watts transmitter power output, 4 dB feedline loss, 3 dB duplexer loss and 6 dB of antenna gain is: 39.7 watts. Add and subtract the dBs to see a total loss of 1 dB. *The answer is slightly less than the original 50 watts.*

The effective radiated power of a repeater with 75 watts transmitter power output, 4 dB feedline loss, 3 dB duplexer and circulator loss and 10 dB of antenna gain is 150 watts. *Solve: Add the dBs and you are positive 3 dB. 3 dB is double, so the effective radiated power would be 150 watts. See how important antenna gain is?*

The effective radiated power of a repeater with 75 watts transmitter power output, 5 dB feedline loss, 4 dB duplexer and circulator loss, and 6 dB of antenna gain is: 37.6 watts. *Solve: You are down a net of 3 dB or half the original power.*

The effective radiated power of a repeater with 100 watts transmitter output, 4 dB feedline loss, 3 dB duplexer and circulator loss and 7 dB of antenna gain is: 100 watts. *Solve: The gain and loss are the same, so the effective output is 100 watts.*

The effective radiated power of a repeater with 100 watts transmitter output, 5 dB feedline loss, 4 dB duplexer and circulator loss and 10 dB antenna gain is: 126 watts. *Solve: You are net 1 dB positive, so the effective output is slightly more than the original 100 watts.*

The effective radiated power of a repeater with 50 watts transmitter power output, 5 dB feedline loss, 4 dB duplexer and circulator loss and 7 dB antenna gain is: 31.5 watts. *Solve: You are down a net of 2 dB, which is almost half. The answer, 31.5 watts, is the only one close.*

SUMMARY Subelement J

TOPIC 63 – ANTENNA THEORY

High SWR caused by detuned antenna coupler.
Radiation resistance is equivalent resistance.

Knowing radiation resistance allows you to match impedances for maximum power transfer.
RF energy is one direction controlled by an isolator.
Trap antenna's advantage is multi-band operation.
Antenna bandwidth is range over which is performs well.

TOPIC 64 – VOLTAGE, CURRENT AND POWER

$P = I^2R$ or $I^2 = P/R$
Voltage in a receive antenna proportional to received field strength.
Best SWR is 1:1.
At ends of antenna, maximum voltage, minimum current.
Attenuation of 1000 times is 30 dB.
30 watts out and 5 reflected is 25 radiated.

TOPIC 65 – FREQUENCY AND BANDWIDTH

Vertical receives equally in all horizontal directions.
Lower resonant frequency (lengthen) with a coil.
A shortened antenna has less bandwidth.
Increase resonant frequency with a capacitor in series.
Antenna produces electrostatic and electromagnetic fields.

TOPIC 66 – TRANSMISSION LINES

Velocity factor is ratio of wave velocity in line to speed of light.
Velocity factor determined by dielectrics of the line.
Nitrogen in lines to prevent moisture.
3 dB is half, 6 dB is one-quarter.
Shorted 1/4 wave stub will cancel even harmonics.
Line loss in dB / 100 feet adds.
10 dB loss passes 1/10 power.

TOPIC 67 – EFFECTIVE RADIATED POWER

Net system losses and gains, then apply dB calculation to input power.

AIRCRAFT (Subelement K)

68 – DISTANCE MEASURING EQUIPMENT

The frequency range of the Distance Measuring Equipment (DME) used to indicate an aircraft's slant range distance to a selected ground-based navigation station is: 962 MHz to 1213 MHz. *Cheat: Airplanes fly high; pick the highest answer.*

The Distance Measuring Equipment (DME) measures the distance from the aircraft to the DME ground station. This is referred to as: the slant range. *Measuring distance is measuring range.*

The Distance Measuring Equipment (DME) ground station has a built-in delay between reception of an interrogation and transmission of the reply to allow: operation at close range. *The equipment needs time to switch from receive to transmit.*

The main underlying operating principle of an aircraft's Distance Measuring Equipment (DME) is: a measurable amount of time is required to receive a radio signal through the Earth's atmosphere. *The DME measures the time it takes for the signal to arrive and calculates distance.*

VORTAC is a radio-based navigation aid. It stands for VHF Omni-directional Range & Tactical Aircraft Control, a navigation beacon.

The navigation aid that determines the distance from an aircraft to a selected VORTAC station by measuring the length of time the radio signal takes to travel to and from the station is: Distance Measuring Equipment. *Distance Measuring Equipment determines distance.*

The majority of airborne Distance Measuring Equipment systems automatically tune their transmitter and receiver frequencies to the paired: VOR/LOC channel. *LOC stands for localizer; the heading information. VOR tells the distance; LOC, the heading.*

69 – VHF OMNIDIRECTIONAL RANGE (VOR)

All directions associated with a VOR are related to: magnetic north. *Magnetic north is easy to determine. Converting to true north would require constant recalculation as you change location.*

The rate that the transmitted VOR variable speed rotates is equivalent to: 30 revolutions per second.

The frequency range of the ground based Very-high-frequency Omnidirectional Range (VOR) stations use for aircraft navigation is: 108 MHz to 117.95 MHz. *That is the only answer in the VHF range (30 – 300 MHz). Watch out for the distractor that is in kHz.*

Lines drawn from the VOR station in a particular magnetic direction are: radials. *The lines radiate out from the VOR station.*

The amplitude modulated variable phase signal and the frequency modulated reference phase signal of a Very-high-frequency Omnidirectional Range (VOR) station used for aircraft navigation are synchronized so that both signals are in phase with each other at: 360 degrees North, magnetic bearing position. *Way too much information! We know the system is based on magnetic North. "360 degrees North, magnetic bearing" is the answer.*

The main underlying operating principle of the Very-high-frequency Omnidirectional Range (VOR) aircraft navigation system is: a phase difference between two AC voltages may be used to determine an aircraft's azimuth position in relation to a selected VOR station. *VOR determines the direction (azimuth) to the VOR station.*

70 – INSTRUMENT LANDING SYSTEM (ILS)

The frequency range of the localizer beam system used by aircraft to find the centerline of a runway during an Instrument Landing System(ILS) approach to an airport is: 108.10 MHz to 111.95 MHz. *108 MHz, same as the VOR station in the previous section.*

The frequency range of the marker beacon system used to indicate an aircraft's position during an Instrument Landing System (ILS) approach to an airport's runway is: 75 MHz. *The answers are much longer, listing marker tones, but look for the one on 75 MHz.*

A required component of an Instrument Landing System (ILS) is the: Localizer: shows the aircraft deviation horizontally from center of the runway. *The other answers look good but are functions of the VORTAC system. ILS is concerned with lining up the runway.*

The antenna used in an aircraft's Instrument Landing System (ILS) glideslope installation is: a folded dipole reception antenna. *The aircraft is receiving ILS information, not transmitting. The antenna is a reception antenna.*

On runway approach, an ILS Localizer shows deviation left or right of runway center line.

The localizer beam system: produces two amplitude modulated antenna patterns; one pattern with an audio frequency of 90 Hz and one pattern with an audio frequency of 150 Hz; one left of the runway centerline and one right of the runway centerline. *The ILS is concerned with left and right. Pick the only answer with left and right.*

71 – AUTOMATIC DIRECTION FINDING EQUIPMENT

The frequency range of an aircraft's Automatic Direction Finding (ADF) equipment is: 190 kHz to 1750 kHz. *ADF operates at low frequencies. Choose the only answer in kHz. The rest are MHz.*

The term "night effect" when using an aircraft's Automatic Direction Finding (ADF) equipment refers to the fact that the: Non Directional Beacon (NDB) can bounce off the Earth's ionosphere at night and be received in almost any direction. *ADF frequencies are in a range that will bounce, and the bounce may skew the direction.*

The transmit and receive frequencies of an aircraft's mode C transponder operating in the Air Traffic Control RADAR Beacon System (ATCRBS) is transmit: at 1090 MHz and receive at 1030 MHZ. *The answer has to be in MHz and the aircraft's RADAR transmits up and listens down.*

In addition to duplicating the function of a mode C transponder, an aircraft's mode S transponder can also provide: mid-air collision avoidance capabilities.

The type of encoding used in an aircraft's mode C transponder transmission to a ground station of the Air Traffic Control, RADAR Beacon System

(ATCRBS) is: pulse position modulation. *RADAR determines position. Pulse position.*

The correct statement about an aircraft's Automatic Direction Finding (ADF) is: an aircraft's ADF antennas can receive transmissions that are over the Earth's horizon (sometimes several hundred miles away) since these signals will follow the curvature of the Earth. *ADF must operate over long distances.*

72 - AIRCRAFT ANTENNA SYSTEMS AND FREQUENCIES

The antenna pattern radiated from a ground station phased-array directional antenna when transmitting the PPM pulses in a Mode S interrogation signal of an aircraft's Traffic alert and Collision Avoidance System (TCAS) installation is: 1030 MHz omnidirectional. *The transmit antenna needs wide coverage, omnidirectional. The listen frequency is 1030 MHz.*

The type of antenna used in an aircraft's Instrument Landing System (ILS) marker beacon installation is: a balanced loop reception antenna. *Loop antennas are directional, and the aircraft needs directions. The antenna is used to receive not radiate.*

The frequency range of an aircraft's Very High Frequency (VHF) communications is: 118 MHz to 136 MHz (worldwide up to 151.975 MHz). *Aircraft operate around the world, so look for the answer that includes "worldwide."*

Aircraft Emergency Locator Transmitters (ELT) operate on: 121.5, 243 and 406 MHz. *Emergency locators need lots of coverage so pick the answer with three frequencies.*

The frequency range of an aircraft's altimeter is: 4250 Mhz to 4350 Mhz. *It is the altimeter so pick the highest frequency answer.*

The type of antenna attached to an aircraft's Mode C transponder installation and used to receive 1030 MHz interrogation signals from the Air Traffic Control Radar Beacon System (ATCRBS) is: an L band monopole blade-type omnidirectional antenna. *Too much information! You want to receive signals from all directions so the antenna should be omnidirectional.*

73 – EQUIPMENT FUNCTIONS

Some aircraft and avionics equipment operates with a prime power line frequency of 400 Hz. The principal advantage of a higher line frequency is: the magnetic devices in a 400 Hz power supply such as transformers, chokes, and filters are smaller and lighter than those used in 60 Hz power supplies. *Smaller and lighter is good in an aircraft.*

Aviation services use predominantly: dynamic microphones. Dynamic microphones are quite sturdy and do not require external power.

Typical airborne HF transmitters usually provide a nominal RF power output to the antenna of: 100 watts, compared with 20 watts RF output from a typical VHF transmitter. *Remember "100 watts HF." There is only one answer with that.*

Before ground testing an aircraft RADAR, the operator should: ensure that the area in front of the antenna is clear of other maintenance personnel to avoid radiation hazards. *RADAR operates at microwave frequencies and can heat body tissue.*

The antenna used in an aircraft's Very High Frequency Omnidirectional Range (VOR) and Localizer (LOC) installations is: horizontally polarized omnidirectional reception antenna. *It is a reception antenna, and it must be omnidirectional.*

The function of a commercial aircraft's SELCAL installation is: a system where a ground-based transmitter can call a selected aircraft or group of aircraft without the flight crew monitoring the ground-station frequency. *SELCAL is "Selective Calling." It is a way of ringing up an aircraft or group of aircraft without constant monitoring. Look for the answer with "group of aircraft."*

SUMMARY Subelement K

TOPIC 68 – DISTANCE MEASURING EQUIPMENT

DME uses 962 MHz – 1213 Mhz.
Distance from aircraft to DME ground station is slant range.
DME has built-in delay to allow for operation at close range.
DME based on time required to receive signal.
VORTAC is a navigation beacon. Distance measured by time to travel.
Tune to the paired VOR/LOC channel.

TOPIC 69 – VHF OMNIDIRECTIONAL RANGE (VOR)

All directions are related to magnetic North.
VOR rotates at 30 revolutions per second.
VOR uses 108 MHz – 117.95 MHz.
Lines from VOR in a particular direction are radials.
Signals in phase at 360 Degrees North.

TOPIC 70 – INSTRUMENT LANDING SYSTEM (ILS)

ILS approach uses 108.10 Mhz – 111.95 MHz.
Marker beacon system to indicate aircraft's position is on 75 MHz.
Required component called Localizer shows deviation right or left from center of runway.
ILS antenna is folded dipole reception antenna.
Localizer produces two audio tones for right and left.

TOPIC 71 – AUTOMATIC DIRECTION FINDING

ADF equipment operates on 190 kHz to 1750 kHz.
Night effect causes signals to bounce off in any direction.
Air Traffic Control RADAR Beacon System (ATCRBS) transmits at 1090 Mhz and receives at 1030 MHz.
Mode S transponder for mid-air collision avoidance.
ATCRBS uses pulse modulation.
ADF signals can follow curvature of the Earth.

TOPIC 72 – AIRCRAFT ANTENNA SYSTEMS AND FREQUENCIES

Traffic alert and Collision avoidance System is on 1030 Mhz with omnidirectional pattern.
ILS uses balanced loop reception antenna.
VHF communications is worldwide.
Emergency Locator Transmitters (ELT) on three frequencies.
Altimeter frequency is 4250 Mhz to 4350 Mhz.
ATCRBS uses omnidirectional antenna.

TOPIC 73 – EQUIPMENT FUNCTIONS

400 Hz power allows use of smaller and lighter transformers.
Microphones are dynamic type.
HF is 100 watts; VHF 20 watts output.
Before testing RADAR make sure no one in front of it.
VOR and LOC use omnidirectional reception antennas.
SELCAL is selective calling. Ring up one or a group of aircraft.

INSTALLATION, MAINTENANCE & REPAIR (Subelement L)

74 – INDICATING METERS

A DMM is a Digital MultiMeter for measuring Volts, Amperes, and Ohms. The digital readout makes it easier to get precise readings.

A ½ digit on a DMM is: partial extended accuracy on the lower side of the range. The first digit on the left side on the screen can be 1, nothing or a – sign.

A 50-microampere meter movement has an internal resistance of 2,000 ohms. The applied voltage to indicate half-scale deflection is: .05 volts. *Solve: E = IR. Multiply current by resistance to get the full-scale deflection in volts. Then, divide by 2 for half-scale. .000050 x 2000 = .1. Divide by 2 = .05 volts.*

The purpose of a series multiplier resistor used in a voltmeter is: to increase the voltage-indicating range of the voltmeter. *It is a multiplier resistor, so it multiplies the range.*

The purpose of a shunt resistor in an ammeter is: to increase the ampere indicating range of the ammeter. *Same principle.*

The instrument used to indicate high and low digital voltage states is: a logic probe. *Logic circuits use High and Low states, so it is only logical to use a logic probe.*

The instrument used to verify proper radio antenna functioning is: an SWR meter. *An SWR meter measures forward and reflected power. High reflected power would indicate a malfunction.*

75- TEST EQUIPMENT

A frequency counter is used: to measure the time between events, or the frequency, which is the reciprocal of the time. *A frequency counter measures the frequency.*

A frequency standard is: a device used to produce a highly accurate reference frequency. *You use a frequency standard as a reference to calibrate a receiver.*

EMI is electromagnetic interference. **The equipment that may be useful to track down EMI aboard a ship or aircraft is: a portable AM receiver.** *You walk around with the receiver hunting for the source.*

On an analog watt meter, the part of the scale which is the most accurate is: full-scale. The upper one-third of the meter is the only true calibrated part. *Full accuracy at full scale.*

The frequency standard used as a time base standard by field technicians is: a Rubidium Standard. A Rubidium oscillator is very accurate.

A multirange AF voltmeter calibrated in dB and a sharp, internal 1000 Hz bandstop filter both used in conjunction with each other to perform quieting tests is: a SINAD meter. *SINAD is SIgnal, Noise, and Distortion. Use a signal, noise and distortion meter for quieting tests.*

76 – OSCILLOSCOPES

To decrease circuit loading when using an oscilloscope:, use a 10:1 divider probe. *The divider probe has a very high impedance and only takes 1/10 of the signal.*

A spectrum analyzer is different from a conventional oscilloscope: The oscilloscope measure electrical signals in the time domain while a spectrum analyzer is used to display electrical signals in the frequency domain. *If you are analyzing a spectrum, you are analyzing frequency.*

The stage that determines the maximum frequency response of an oscilloscope is: the vertical amplifier. *The vertical rise time must be short enough to accommodate a higher frequency.*

The factors that limit the accuracy, frequency response and stability of an oscilloscope are: sweep oscillator quality and deflection amplifier bandwidth. *Quality and bandwidth limit accuracy.*

An oscilloscope can do all of the following except: measure velocity of light with the aid of a light emitting diode. *An oscilloscope can't measure the speed of light.*

The instrument used to check the signal quality of a single-sideband radio transmission is: an oscilloscope. *The oscilloscope can examine the output waveform for distortion.*

77 – SPECIALIZED INSTRUMENTS

To measure the characteristics of a transmission lines use a time-domain reflectometer (TDR) consisting of: an oscilloscope and pulse generator. The pulse generator sends out a pulse, and the oscilloscope analyzes the return signal.

The horizontal axis of a spectrum analyzer displays: frequency. *You see the spectrum of frequencies.*

The vertical axis of a spectrum analyzer displays: amplitude. Taller spike, stronger signal.

The instrument most accurate when checking antennas and transmission lines at the operating frequency of the antenna is: a frequency domain reflectometer. *If you are checking at the operating frequency, use a frequency domain reflectometer.*

The test instrument used to display spurious signals in the output of a radio transmitter is: a spectrum analyzer. *You are analyzing the output spectrum.*

The instrument commonly used by radio service technicians to monitor frequency, modulation, check receiver sensitivity, distortion and generate audio tones is: a service monitor. *Service technicians use a service monitor to do many things.*

78 – MEASUREMENT PROCEDURES

P25 is a standard for digital communications.
A P25 radio system can be monitored with a scanner: if the scanner has P25 decoding. *Makes sense the scanner needs the ability to decode the digital signals.*

It is true that: the RF power reading on a CDMA (Code Division Multiple Access) radio is not accurate on an analog power meter. *Code Division Multiple Access has pulses and a swinging-needle analog meter can't keep up.*

A common method to program radios without using a "wired" connection is: infra-red communication. The radios connect and communicate using infra-red light pulses.

The common method for determining the exact sensitivity specification of a receiver: measures the recovered audio for 12 dB of SINAD. SINAD is signal, noise, and distortion. *The standard is a dozen.*

A communication technician would perform a modulation-acceptance bandwidth test to: determine the effective bandwidth of a communications receiver. *A modulation-acceptance bandwidth test tests bandwidth.*

The maximum FM deviation for voice operation of a normal wideband channel on VHF and UHF is: 5 kHz. Deviation is the frequency sway away from the center frequency and 5 kHz is the standard for wideband FM.

79 – REPAIR PROCEDURES

When soldering or working with CMOS electronics products or equipment, a wrist strap: must have less than 100,000 ohms of resistance to prevent static electricity. *The wrist strap is to prevent the build-up of static electricity that could damage the CMOS*[14] *device. Don't bother memorizing the number of ohms.*

The preferred method of cleaning solder from plated-through circuit board holes is to: use a vacuum device. *The vacuum sucks the solder out of the holes without damaging the board.*

The proper way to cut plastic wire ties is: with flush-cut diagonal pliers and cut flush. *You don't want to leave a sharp edge sticking out to snag someone.*

The ideal method of removing insulation from wire is: the thermal stripper. *A thermal stripper melts the insulation off whereas, knives and other stripping tools can nick and weaken the wire.*

A "hot gas bonder" is used: to allow non-contact melting of solder. *Hot gas doesn't require a physical contact with the part or board.*

When repairing circuit board assemblies, it is most important to: wear safety goggles. *Safety glasses are always a good idea. You don't want a blob of hot solder or a wire shard in your eye.*

---

[14] Complimentary Metal-Oxide Semiconductor. This is a precaution in addition to the Zener Diode in the CMOS to prevent zapping.

80 – INSTALLATION CODES & PROCEDURES

The color of the binder for pairs 51-75 in a 100 pair cable is: green. There aren't 100 different colors to use on individual wires. They are bundled in groups of 25 and wrapped in a colored binder. 51-75 is in a green binder. *Cheat: Go with the green.*

The coding goes even deeper. **The 6th pair color code in a 25 pair switchboard cable as is found in building and telecommunications interconnections is: Red/Blue, Blue/Red.** *Cheat: Stick with the red, white and blue.*

When routing cables in a mobile unit, it is most important: to assure radio and electronics cables do not interfere with the normal operation of the vehicle.

You should not use white or translucent plastic tie wraps on a radio tower because: UV radiation from the Sun deteriorates the plastic very quickly. *White and translucent ties are not UV resistant. Usually, black ties are UV resistant.*

The tolerance off plumb for installing a single base station radio rack is: just inside the bubble level. *That would be "level."*

The type of wire connecting an SSB automatic tuner to an insulated backstay would be: GTO-15 high-voltage cable. *Antenna tuners can generate high voltage, so you want a high-voltage cable. Look for the high-voltage answer.*

81 – TROUBLESHOOTING

On a 150 watt marine SSB HF transceiver, a steady output of 75 watts would indicate: there is probably a defect in the system causing the carrier to be transmitted. *On SSB, there should be no steady output unless feeding the radio a tone. Suspect something is triggering the carrier.*

The tachometer of a buildings elevator circuit experiences interference caused by the radio system. A common potential "fix" would be to: add a .01 μF capacitor across the motor/tachometer leads. *A capacitor shorts out the RF interference. .01 μF is all you need.*

The common method of programming portable or mobile radios is to use a: laptop computer. *Program using a computer program.*

In a software-defined transceiver, the best way for a technician to make a quick overall evaluation of a radio's operational condition would be to: use the built-in self-test feature. *A built-in self-test feature would be easy to use.*

An installer might verify correct GPS sentence to marine DSC[15] VHF radio by: looking for latitude and longitude on the display. The GPS "sentence" is the message: latitude, longitude, speed, bearing, time, etc. *Look at the display makes sense.*

To activate the DSC emergency signaling function on a marine VHF one must: input a registered 9-digit MMSI. *The MMSI is the Maritime Mobile Service Identity, a unique 9-digit number identifying your radio.*

---

[15] Digital Selective Calling

ELEMENT 3 – INSTALLATION MAINTENANCE & REPAIR

SUMMARY Subelement L

TOPIC 74 – INSTALLATION, MAINTENANCE & REPAIR

½ digit on a DMM is partial extended accuracy.
Meter deflection is E = IR
Series multiplier resistor increases range of voltmeter.
Shunt resistor increases range of ammeter.
Measure high and low voltage states with logic probe.
Verify proper function of antenna with SWR meter.

TOPIC 75 – TEST EQUIPMENT

Frequency counter measures time between events.
Frequency standard produces accurate reference frequency.
Track down EMI with portable AM receiver.
Analog watt meter is most accurate at full scale.
Frequency standard based on Rubidium.
Perform quieting tests with a SINAD meter.

TOPIC 76 – OSCILLOSCOPES

Decrease circuit loading with a 10:1 probe.
Oscilloscope measures in time domain. Spectrum analyzer measures in frequency domain.
Max frequency response limited by vertical amplifier.
Oscilloscope cannot measure the speed of light.
Check SSB signal quality with an oscilloscope.

TOPIC 77 – SPECIALIZED INSTRUMENTS

Measure characteristics of transmission lines with time-domain reflectometer. (Oscilloscope and pulse generator).
Horizontal axis of spectrum analyzer is frequency.
Vertical axis of spectrum analyzer is amplitude.
Display spurious signals with a spectrum analyzer.
Use a service monitor to monitor frequency, modulation, sensitivity, distortion, audio tones.

TOPIC 78 – MEASUREMENT PROCEDURES

P25 scanner needs P25 decoding.
RF power reading on CDMA (Code Division Multiple Access) is not accurate on analog meter.
Program radios with infra-red communication.
Sensitivity measures audio for 12 dB of SINAD.
Modulation bandwidth test determines bandwidth.
Maximum wideband FM deviation is 5 kHz.

TOPIC 79 – REPAIR PROCEDURES

Use a wrist strap to prevent static electricity.
Clean solder from boards with a vacuum device.
Cut plastic ties flush with flush-cut diagonal pliers.
Remove insulation with thermal stripper.
Hot gas bonder allows non-contact melting of solder.
When repairing, use safety goggles.

TOPIC 80 – INSTALLATION CODES AND PROCEDURES

Pairs 51-75 are in a green binder.
Sixth pair color is Red/Blue Blue/Red.
Route cables to not interfere with vehicle operation.
White or translucent ties deteriorate in sun.
Plumb is just inside the bubble.
Wire to insulated backstay with high-voltage cable.

TOPIC 81 – TROUBLESHOOTING

75 watts steady output in 150 watt SSB transmitter indicates defect causing carrier to be transmitted.
Install .01 µF capacitor across leads to stop RFI.
Program radios with laptop computer.
Quick over-all evaluation with self-test feature.
Verify correct GPS sentence by looking for lat/long on display.
To activate DSC emergency signaling function input 9 digit MMSI.

COMMUNICATIONS TECHNOLOGY
(Subelement M)

82 – TYPES OF TRANSMISSIONS

The term describing a wide-bandwidth communications system in which the RF carrier frequency varies according to a pre-determined sequence is: spread spectrum communication. *You've seen this question before. If the carrier frequency varies in a pre-determined sequence, it is spread spectrum.*

The two types of spread spectrum systems used in most RF communications applications are: direct sequence and frequency hopping. *"Sequence" and "hopping" define spread spectrum/*

The term used to describe spread spectrum communications where the center frequency of a conventional carrier is altered many times per second in accordance with a pseudo-random list of channels is: frequency hopping. *Don't over-complicate this. If the carrier frequency is altered, it is frequency hopping.*

A TDMA radio carries multiple conversations sequentially in: separate time slots. *TDMA is Time Division Multiple Access. Multiple separate time slots. Another hint is the word "sequentially."*

SSB voice transmissions normally use: J3E emissions, which consist of one sideband and a suppressed carrier. *SSB is Single Sideband, one sideband.*

The two most-used PCS (Personal Communications Systems) coding techniques used to separate different calls are: CDMA and GSM. *CDMA is Code Division Multiple Access, a* *coding technique. If you see CDMA, that is the answer.*

83 – CODING AND MULTIPLEXING

CODEC is: a coder/decoder IC or circuitry that converts a voice signal into a predetermined digital format for encrypted transmission. *Too complicated. Recognize CODEC is Coder/Decoder and look for that in an answer.*

The type of CODEC used by the GSM (Global System for Mobile Communications) is called: Regular-Pulse Excited (RPE). *Get excited about being RoPEd into global communications.*

The code gaining the widest acceptance for exchange of data from one computer to another is: ASCII.

The International Organization for Standardization has developed a seven-level reference model for packet-radio communications structure. The level responsible for the actual transmission of data and handshaking signals is: the physical layer. *Handshaking is physical.*

The level that arranges the bits into frames and controls data flow is: the link layer. *Data flow is linked.*

The CODEC used in Phase 2 P25 radios is: AMBE.

84 – SIGNAL PROCESSING, SOFTWARE AND CODES

An SDR is: a software defined radio.

DSP[16] in a modern DSP radio does not control: SWR. *SWR is a function of antenna and feedline mismatch and has nothing to do with digital signal processing.*

The code used for GMDSS-DSC[17] transmissions is: a 10-bit error correcting code starting with bits of data followed by a 3-bit error correcting code. *Remember 10 bits followed by 3.*

The code used for SITOR[18]-A and –B transmissions is: each character consists of 7 bits with 4 "zeroes" and "3 "ones." *Remember 4 zeroes.*

The Idle signal (a) (0000111) is used for FEC Phasing Signal 1. *Cheat: I, (a), 1 are all for Signal 1.*

The principle that allows multiple conversations to be able to share one radio channel on a GSM channel is: time division multiplex. Conversations are broken into bursts sent at different times. TDMA was also covered in topic 82.

SUMMARY Subelement M

TOPIC 82 - TYPES OF TRANSMISSIONS

RF carrier frequency varies in pre-determined sequence with spread-spectrum.
Two types of spread spectrum: direct sequence and frequency hopping.

---

[16] Digital Signal Processing
[17] Global Marine Distress and Safety System – Digital Selective Calling
[18] Simplex Telex Over Radio

Center channel altered in pseudo-random list of channels is frequency hopping.
SSB voice is J3E, one sideband and suppressed carrier.
PCS (Personal Communication Systems) use CDMA and GSM coding.

TOPIC 83 – CODING AND MULTIPLEXING
CODEC is coder/decoder converts voice to digital.
CODEC used by GSM (Global System for Mobile Communications) is Regular-Pulse Excited (RPE).
Exchange of data between computers uses ASCII.
Transmission of data and handshaking is the physical layer.
Link layer arranges bits into frames and controls data flow.
CODEC in Phase 2 P25 radios is AMBE.

TOPIC 84 – SIGNAL PROCESSING, SOFTWARE AND CODES
DSP (Digital Signal Processing) does not control SWR.
GMDSS-DSC uses 10-bit error correction, bits of data followed by 3-bit error code.
SITOR (Simplex Telex Over Radio) uses 7 bits with 4 zeroes.
Idle signal (a) 0000111 used for FET Phasing Signal 1.
Multiple conversations share one channel on GSM (Global System for Mobile Communication) through time division multiplex.

MARINE (Subelement N)

85- VHF

The channel spacing used for VHF marine radio is: 25 kHz. *You want room between channels so pick the answer with the widest spacing*

The VHF channel assigned for distress and calling is: 16.

The VHF channel used for Digital Selective Calling and acknowledgment is: 70.

The maximum allowable frequency deviation for VHF marine radios is: +/- 5 kHz.

The reason for the USA-INT control or function is: to change some channels that are normally duplex to simplex. *Look for the answer that changes "some" channels. Not all channels need changing.*

An installer might verify correct GPS sentence to marine DSC VHF radio: by looking for latitude and longitude, plus speed, on VHF display. *Comparing the readouts.* Recognize the "GPS sentence" from Topic 82.

86 – MF-HF, SSB-SITOR

A common occurrence when voice-testing an SSB aboard a boat is: voltage panel indicator lights may glow with each syllable. *RF is getting into the wires.*

Apparent low voltage on marine SSB transmitting might be: a blown black negative fuse. The negative side is no longer directly connected to the power source and current runs through the radio chassis and other routes. Those routes may have a higher resistance which causes a voltage drop.

The wire that connects an SSB automatic antenna tuner to an insulated backstay is: GTO-15 high-voltage cable. *We've seen this before. The antenna tuner can generate high voltages, so you need high-voltage cable. You don't need to know it is GTO-15.*

With SITOR communications: ARQ[19] message transmissions are made in data groups consisting of three-character blocks. *SITOR is Simplex Teletype Over Radio, a data mode, so the transmission is in data groups. Look for the answer with "data groups."*

ARQ, FEC and SFEC refer to: two-way communications, one-way communications to all stations, one-way communication to single stations. *ARQ, automatic repeat request, is between two stations. The sender waits for confirmation from the receiving station before continuing. FEC, forward error correction, repeats the sending data. FEC is one-way and can go to many stations because it does not rely on a confirmation. SFEC is Super Forward Error Correction to a single station. Look for 2-1-1.*

With SITOR: two-way communication with the coast station using FEC is not necessary to be able to receive broadcasts. *FEC is forward error correction. It repeats data and does not rely on feedback from the receiving station.* When sending using FEC, you do not need to receive confirmations to continue the message.

---

[19] Automatic Repeat ReQuest

87- SURVIVAL RAFT EQUIPMENT: VHF, SARTs & EPIRBs

A SART[20] begins transmission: after being activated and responding to a RADAR interrogation. It remains in standby mode to preserve battery until it senses nearby RADAR.

The signal from a Search and Rescue Transponder will appear on a RADAR display: as a series of 12 equally spaced dots. The dots point to the SART.

A search and rescue transponder operates in the: 9 GHz band. RADAR X band.

The piece of required GMDSS equipment that is the primary source of transmitting locating signals is: an EPIRB[21] transmitting on 406 MHz. *The important take-away in the answer is "EPIRB" transmitting your position. The frequency is not an issue.*

Once activated, EPIRBs transmit a signal for use in identifying the vessel and for determining the position of the beacon.

406 MHz EPIRB transmissions: transmit a unique hexadecimal identification number. *The unique number is how the EPRIB identifies the vessel.*

---

20 Search And Rescue Transponder

21 Emergency Position Indicating Radio Beacon

88- FAX, NAVTEX

Facsimile is: transmission of printed pictures for permanent display on paper. *We all know a fax is on paper.*

The standard scan rate for high-frequency 3 Mhz – 23 Mhz weather facsimile reception from shore stations is: 120 lines per minute. Weather fax is slow but not the slowest answer.

The bandwidth of a good crystal lattice band-pass filter for weather facsimile HF reception would be: 1 kHz at -6 dB. *Remember 1 kHz, the -6 dB is extra information.*

NAVTEX[22] receives MSI[23] broadcasts using SITOR-B or FEC mode. *NAVTEX receives MSI broadcasts. Broadcasts use forward error correction, FEC, because the receiver is not expected to respond. Look for FEC in the answer.*

The primary frequency used exclusively for NAVTEX broadcast is: 518 kHz. Just below the AM radio dial.

Whether a NAVTEX receiver does not print a particular type of message is determined by: whether the subject indicator matches that programmed for rejection by the operator. *If the number has been blocked, you won't get the call.*

---

22 Navigational Telex
23 Marine Safety Information

89- NMEA Data

The data language that is bi-directional, multi-transmitter, multi-receiver network is: NMEA[24] 2000. *Cheat: Bi-directional is 2 way. Look for 2000 in the answer.*

Shielding on an NMEA 0183 data line should be: terminated to ground at the talker end and unterminated at the listener. *Unterminating at the listener end is a way to avoid ground loops.*

In NMEA network topology: if one device should fail, there will be no interruption to the other devices. *That is good network design.*

In an NMEA 2000 device, a load equivalence number (LEN) of 1 is equivalent to: 50 mA current consumption.

An NMEA 2000 system with devices in a single location may be powered by: an end-powered network. *The power supply may be at one end.*

The voltage drop at the end of the last segment that satisfies NMEA 2000 network cabling plans is: 1.5 volts. Any larger drop and the downstream devices may not function properly.

SUMMARY Subelement N

TOPIC 85 - VHF

Channel spacing for VHF marine is 25 kHz.
Channel for distress and calling is 16.
Digital Selective Calling on channel 70.
Maximum FM frequency deviation is +/- 5 kHz.
USA-INT control changes some channels that are normally duplex to simplex.

[24] National Marine Electronics Association

Check correct GPS sentence on DCS VHF radio by looking for lat/lon and speed on VHF display.

Topic 86 – MF-HF, SSB-SITOR

Testing SSB may cause panel lights to glow.
Low voltage on transmitting might be blown negative fuse.
Connect to backstay with high-voltage cable.
SITOR (Simplex Teletype Over Radio) communications are in data groups.
ARQ is two-way communications, FEC and SFEC are one-way communications. (Forward Error Correction)
SITOR using FEC does not require two-way communications.

Topic 87 – SURVIVAL RAFT EQUIPMENT, VHF, SARTS & EPIRBS

SART begins after being activated by RADAR.
SART signals are 12 equally spaced dots.
SART operates on 9 GHz RADAR band.
Primary GMDSS equipment is an EPIRB.
EPRIB signal identifies the vessel and is a beacon.
EPRIB transmits unique hexadecimal ID number.

TOPIC 88 – FAX, NAVTEX

Weather fax at 120 lines per minute.
Bandwidth for filter 1 kHz.
NAVTEX receives MSI broadcast using SITROR or FEC mode.
NAVTEX on 518 kHz.
NAVTEX skips messages if programmed for rejection.

TOPIC 89 – NMEA data

Bi-directional, multi tx/rx network is NMEA 2000.
NMEA 0183 shielding terminated to ground at one end.
NMEA failure of one device should not interrupt others
NMEA load equivalence number oxc f 1 is 50 mA
Voltage drop at end of NMEA 2000 network not to exceed 1.5 volts

RADAR (Subelement O)

90 – RADAR THEORY

The normal pulse repetition rate is 500 – 2,000 pulses per second (pps).

The RADAR range in nautical miles to an object can be found by measuring the elapsed time during a RADAR pulse and dividing this by: 12.346 µs. *Cheat: Count 12.34.* RADAR takes 6.173 microseconds to travel one nautical mile. One mile out and one mile back is 12.346.

The normal range of pulse widths is: .05 µs to 1.0 µs.

Shipboard RADAR is most commonly operated: in the SHF band. Super High Frequency.

The pulse repetition rate (prr) of a RADAR refers to the: pulse rate of the magnetron. *Pulses originate in the magnetron.* A magnetron is a high-powered vacuum tube that generates microwaves using the interaction of a stream of electrons with a magnetic field while moving past a series of open metal cavities.

If the elapsed time for a RADAR echo is 62 microseconds, the distance to the object is: 5 nautical miles. *Solve: 62 / 12.346 = 5.*

91 - COMPONENTS

The ATR box: prevents received signal from entering the transmitter. ATR is Anti Transmit Receive. It disconnects the transmitter during the receive cycle. *It is an anti-transmit box, so it disconnects the transmitter.*

The purpose or function of the RADAR duplexer/circulator is an: electronic switch that

allows the use of one antenna for both transmission and reception. *A duplexer allows double use for the antenna.*

To determine the performance of a RADAR system at sea: use an echo box. At sea, there may be nothing for the RADAR to "see." *The echo box creates an echo.*

The purpose of a synchro transmitter and receiver is: to transmit the angular position of the antenna to the indicator unit. *The antenna and indicator must line up. Don't fall for the answer that has the word "synchronize."*

Digital signal processing (DSP) of RADAR signals (compared with analog): causes improved weak signal or target enhancement. *DSP removes noise.*

The component or circuit providing the transmitter output power for a RADAR system is the: magnetron.

92 – RANGE, PULSE WIDTH & REPETITION RATE

When a RADAR is being operated on the 48-mile range setting, the most appropriate pulse width (PW) and pulse repetition rate (pps) is 1.0 µs PW and 500 pps. PPS is pulses per second and the only focus of this question. You have to wait for a pulse to come back before sending another one. *Solve: The speed of light is 186,282 miles per second. To travel 96 miles (out and back) takes 48 / 186,282 = .00051 seconds. 1 pulse every .00051 seconds would be 1 / .00051 = 1,960 pulses per second. There is only one answer with less than 1,960 pps. Cheat: 48 miles is a long way. Pick the answer with the fewest pps.*

For a target 25 miles away, when the RADAR is being operated on the 25 miles range, the most appropriate pulse width and pulse repetition rate is: 1.0 µs and 500 pps. *Cheat: Same answer as above 1.0 µs and 500 pps.*

When a RADAR is being operated on the 6-mile range setting, the most appropriate pulse width and pulse repetition rate is: 0.25 µs PW and 1,000 pps. *Cheat: You could do the math but the answer is subjective — Pick 1,000 pps as a mid-range of the answers for the nearby target.*

Then RADAR is being operated on the 1.5 miles range setting the most appropriate pulse width, and pulse repetition rate is: .05 µs and 2,000 pps. *Up close, it is short pulses and lots of them.*

The pulse width and repetition rate to use at long ranges is: wide pulse and slow repetition. *Wider pulse to get a stronger return and that implies a slower repetition.*

The pulse width and repetition rate to use at short ranges is narrow pulse and fast repetition rate. *Narrow pulse is all you need, and that implies a quicker repetition.*

93 – ANTENNAS AND WAVEGUIDES

When the frequency doubles, the gain of a parabolic dish antenna changes by: 6 dB. Doubling the frequency increases the gain 4 times.

The type of antenna or pickup device used to extract the RADAR signal from the wave guide is a: J-hook. *Cheat: J looks like a hook. Imagine a J-hook extracting the signal. If you see "J-hook" in an answer, it is correct.*

As gain is increased: the beamwidth of an antenna decreases as the gain is increased. The antenna radiates the same amount of energy. Gain is increased by narrowing the beam and concentrating that energy.

A common shipboard RADAR antenna is the: slotted array.

Conductance takes place in a waveguide: through electromagnetic and electrostatic fields in the walls of the waveguide." *The two fields guide the wave.*

To couple energy into and out of a waveguide use a thin piece of wire as an antenna. *Look for a wire antenna in the answer.*

94 – RADAR EQUIPMENT

The permanent magnetic field that surrounds a traveling-wave tube (TWT) is intended to: prevent the electron beam from spreading. The magnetic field concentrates the beam.

Prior to testing any RADAR system, the operator should first: assure no personnel are in front of the antenna. RADAR operates at microwave frequencies and can heat like a microwave oven.

In the term "ARPA RADAR," ARPA is the acronym for: Automatic RADAR Plotting Aid.

To ensure the magnetron is not weakened, you do not need to: keep the TR properly tuned. *The question asks what is NOT needed. The TR switch is not relevant.*

Exposure to microwave energy from RADAR or other electronic devices is limited to: 5 mW per centimeter. *Cheat: RADAR has five letters.*

RADAR collision avoidance systems do not utilize: your ship's anemometer. *Anemometers measure wind speed not direction, speed or position.*

SUMMARY Subelement O

TOPIC 90 – RADAR THEORY

Normal pulse repetition is 500 – 2000 per second.
Elapsed time / 12.346 = distance.
Normal pulse width is .05 μs to 1.0 μs.
RADAR in the SHF band. (Super high frequency)
Pulse repetition rate is pulse rate of magnetron.

TOPIC 91 – COMPONENTS

ATR box prevents received signal from entering transmitter.
RADAR duplexer/circulator allows use of one antenna for both transmit and receive.
To test at sea, use an echo box.
Synchro transmitter matches antenna position to indicator.
DSP of RADAR causes improved weak-signal and target enhancement.
Transmitter output power is from magnetron.

TOPIC 92 – RANGE, PULSE WIDTH, REPETITION RATE

48-mile range: width 1.0 μs, repetition 500 pps.
25 mile range: 1.0 μs, repetition 500 pps.
6 mile range : .25 μs, repetition 1,000 pps
1.5 mile range: .05 μs, repetition 2,000 pps
Long range use wide pulse and slow repetition.
Short range use narrow pulse and fast repetition.

TOPIC 93 – ANTENNAS AND WAVEGUIDES

If frequency doubles, gain of parabolic dish increases 6 dB.
Extract RADAR signal from wave guide with a J-hook.
As antenna gain increases, beamwidth decreases.
Shipboard RADAR antenna is slotted srray.
Conductance through a waveguide controlled by electromagnetic and electrostatic fields.
Couple energy in and out of a waveguide with a thin piece of wire as an antenna.

TOPIC 84 – RADAR EQUIPMENT

Magnetic field surrounding a traveling-wave tube prevents electron beam from spreading.
Prior to testing RADAR make sure no one in front of antenna.
ARPA RADAR is Automatic RADAR Plotting Aid.
To ensure magnetron is not weakened, you do NOT need to keep TR properly tuned.
Exposure to microwave energy limited to 5 mW per centimeter.
RADAR collision avoidance systems do not use ship's anemometer (wind speed gauge).

SATELLITE (Subelement P)

95 –LOW EARTH ORBIT SYSTEMS

The orbiting altitude of the Iridium satellite communications system is: 485 miles. *Iridium satellites communicate with a handheld transceiver. They are low orbit.*

The frequency band used by the Iridium system is: 1616 – 1626 Mhz. *Cheat: Iridium satellite has 16 letters.*

The services provided by the Iridium system are: digital voice and data at 2.4 kbps. *Digital, of course. Slow speed.*

The Iridium system has 48 spot beams per satellite with a footprint of 30 miles in diameter. *Remember, a footprint 30 miles in diameter.*

The main function of the COSPAS-SARSAT satellite system is: to monitor 406 MHz for distress calls from EPIRBs. *COSPAS is Russian for "space system for search of vessels in distress." The SAR in SARSAT is "search and rescue." Register your EPIRB with the system and the system will track you in a disaster. Memorizing the frequency is not necessary.*

The COSPAS-SARSAT satellite system determines the position of a ship in distress: by measuring the Doppler shift of the 406 MHz signal taken at several different points in its orbit.

96 – INMARSAT COMMUNICATIONS SYSTEMS – 1

INMARSAT is a satellite communication company providing telephone and data services using 13 geosynchronous satellites. **The Orbital altitude of INMARSAT Satellites is: 22,177 miles.** *INMARSAT satellites are geosynchronous; they stay in one spot in the sky. To do that, they hover at 22,177 miles.*

The following describes the INMARSAT system: AOR-E at 15.5° W, AOR W at 54° W. POR at 178° E and IOR at 64.5 ° E. *Mercy! The system covers the whole earth. The AOR-E satellite (Atlantic Ocean Region East) hangs at 15.5° west of 0° longitude to cover the Atlantic Ocean. Pick the answer at 15.5° W, just west of the prime meridian, to cover the Atlantic and ignore the rest of the answer.*

The directional characteristics of the INMARSAT-C SES antenna are: omnidirectional. SES is "ship earth station." *Think of "C" as in "compact. The antenna is omnidirectional for maximum coverage without turning mechanisms.*

When engaging in voice communications via an INMARSAT-B terminal, the technique used is: CODECs digitize the voice signal. *INMARSAT sends data, and a CODEC converts voice to data.*

INMARSAT geostationary satellites provide coverage to vessels in nearly all of the world's navigable waters. *INMARSAT means International Satellite Marine Organization. It covers nearly all of the world except for the polar regions.*

The conditions that can render INMARSAT-B communication impossible include:

- **An obstruction, such as a mast, causing disruption of the signal**
- **A satellite whose signal is on a low elevation, below the horizon**
- **Travel beyond the effective radius of the satellite.**
- **All of these.**

97 – INMARSAT COMMUNICATIONS SYSTEMS – 2

The best description of the INMARSAT-C system is: a store-and-forward system that provides routine and distress communications. *"Store-and-forward" "routine and distress."*

The INMARSAT mini-M system is a: satellite system using spot beams to provide for small craft communications. *Mini system for small craft.*

INMARSAT-B services are: voice at 16 kbps, Fax at 14.4 kbps and high-speed Data at 64/54. *Recognize the answer with "voice, fax, and data at certain kbps," not kHz.*

The INMARSAT systems that offer High-Speed Data at 64/54 kbps are: B, M4, and Fleet. *Look for the answer with "fleet," as in "fleet of foot," meaning "High-Speed."*

When INMARSAT-B and INMARSAT-C (compact) terminals are compared: INMARSAT-C antennas are small and omni-directional, while INMARSAT-B antennas are larger and directional. *"C" for "compact." Don't fall for the answer that suggests a parabolic antenna for lower gain. A parabolic antenna would have higher gain.*

The service provided by INMARSAT-M is: voice at 6.2 kbps, Data at 2.4 kbps, Fax at 2.4 kbps and e-mail. *The standard Voice, Data, Fax plus –M adds eMail. Don't memorize the speeds.*

98- GPS

Global Positioning Service (GPS) satellite orbiting altitude is: 12,554 miles.

The GPS transmitted frequencies are: 1227.6 MHz and 1575.4 MHz.

There are normally: 24 GPS satellites in operation.

GPS satellite orbits are: in six orbital planes equally spaced and inclined about 55 degrees to the equator. *Remember "six orbital planes" or "55 degrees." Either gets you the correct answer.*

To provide complete position and time: 4 satellites must be received.

DGPS is: a system to provide additional correction factors to improve position accuracy. *DGPS stands for Differential Global Positioning System.*

SUMMARY Subelement P

TOPIC 95 – LOW EARTH ORBIT SYSTEMS

Iridium satellite orbit at 485 miles.
Iridium frequency 1616 – 1626 MHz.
Iridium provides digital voice and data at 2.4 kbps.
Iridium has footprint diameter of 30 miles.
COSPAS-SARSAT monitors for EPRIB distress calls.
COSPAS-SARSAT determines position by measuring Doppler shift.

Topic 96 AND 97 – INMARSAT COMMUNICATIONS SYSTEMS

INMARSAT orbit at 22,177 miles.
INMARSAT at AOR-E at 15.5ºW ...
INMARSAT-C SES (Ship Earth Station) antenna is omnidirectional.
Voice communications use CODECs to digitize voice.
INMARSAT geostationary satellites cover nearly all Earth's navigable waters.
May be blocked by obstruction, low satellite elevation, travel beyond the effective range. (All of the above).
INMARSAT–C is store and forward system for routine and distress communications.
INMARSAT mini-M system for small craft.
INMARSAT-B is voice fax and data at X kbps.
INMARSAT high-speed data is B, M4 and Fleet.
INMARSAT-C antennas are small and omni-directional,
INMARSAT-B antennas are larger and directional.
INMARSAT-M adds eMail.

TOPIC 98 – GPS

GPS satellites at 12,554 miles
Transmit on 1227.6 MHz and 1575.4 MHz.
Normally 24 satellites in operation in six orbital planes.
To provide complete position and time requires 4 satellites.
DGPS (Differential GPS) provides additional correction factors.

SAFETY (Subelement Q)

99 - RADIATION EXPOSURE

Compliance with MPE," or Maximum Permissible exposure to RF levels for "controlled" environments, are averaged over: 6 minutes while "uncontrolled" RF environments are averaged over 30 minutes. "Controlled" refers to where you control of the RF exposure. "Uncontrolled" means the public, who is often unaware. *Averaging over a long period gives the public more protection so pick the answer with the longest time for "uncontrolled."*

Sites having multiple transmitting antennas must include antennas with more than: 5% of the maximum permissible power density exposure limit when evaluating RF site exposure.

RF exposure from portable transceivers may be harmful to the eyes because: RF heating may cause cataracts.

The aggregate power level requiring an MPE (Maximum Permissible Power Exposure) study is: 1000 Watts ERP. ERP is "effective radiated power" combining the transmitter power and any antenna gain. *1000 watts.*

If the MPE (Maximum Permissible Exposure) power is present, personnel accessing the affected areas must be trained and certified: yearly. *Think of annual safety inspections.*

You must never look directly into a fiber optic cable because: an active fiber signal may burn the retina and the infra-red light cannot be seen.

100 – SAFETY STEPS

The device that can protect a transmitting station from a direct lightning hit is: "There is no device to protect a station from a direct hit from lightning." We can attempt to bleed off static but, there is no device that can protect a station from a direct hit. Thor's hammer is too powerful.

A shorted-stub lightning protector will only work on the tuned frequency band.

The reason not to put sharp corners on ground leads within a building is: lightning will jump off the ground lead because it is not able to make sharp bends. Lightning is a bull-in-a-china-shop charging straight ahead. It will keep going straight if there is a sharp corner.

You should not use a drill without eye protection.

The class of fire that is caused by an electrical short circuit and the preferred substance used to extinguish that fire is: FE30.

A GFI electrical socket is used: to prevent electrical shock by sensing ground path current and shutting the circuit down. *GFI stands for "Ground Fault Interrupter."*

SUMMARY Subelement Q

Topic 99 – RADIATION EXPOSURE

MPE (Maximum Permissible Exposure) 6 minutes "controlled" 30 minutes "uncontrolled."
All antennas with more than 5% included.
RF from portable transceiver may harm eyes because RF heating causes cataracts.
MPE study required at 1000 watts.
Personnel must be trained and certified yearly.
Never look in fiber optic cable, may burn retina.

TOPIC 100 – SAFETY STEPS

No device can protect against direct lightning strike.
Shorted-stub will only work at tuned frequency.
No sharp corners on ground leads as lightning does not make sharp bends.
Do not use a drill without eye protection.
Electrical fire is FE30.
GFI socket prevents shock by sensing ground path current and shutting off.

SATELLITES, SERVICES & FREQUENCIES

Here are the satellites, services and frequencies that may be on your test.

SERVICE	FREQUENCY
ADF[8]	190 kHz – 1750 kHz
NAVTEX[6]	518 kHz
Watch	2182 kHz, Ch-16
DSC Watch[1]	8 MHz and Ch-70
ILS[3]	75 Mhz
ILS localizer[4]	108.1 - 111.95 MHz
VOR[5]	108 MHz – 117Mhz
Aircraft	121.500 MHz A3E
ELT[7]	121.5, 243, 406 MHz
VHF Radio	151.975 MHz Worldwide
DME[12]	962 MHz – 1213 MHz
RADAR Beacon	1030 MHz Receive
RADAR Beacon	1090 MHz Transmit
GPS[9]	1227.6 and 1575.4 MHz
Iridium[10]	1616 MHz – 1626 MHz
Altimeter	4250 MHz – 4350 MHz
SART X-Band Radar[2]	9 GHz
INMARSAT[11]	Not on test

[1] Digital Selective Calling
[2] Search and Rescue Transponder
[3] Instrument Landing System
[4] Signals left or right of runway.
[5] VHF Omnidirectional Range
[6] Navigational TELEX
[7] Emergency Locator Transmitter
[8] Automatic Direction Finding
[9] Global Positioning Satellite. Alt 12,554 miles
[10] Digital voice and data satellite. Alt 485 miles
[11] Two-way data and messaging. Alt 21,177 miles
[12] Distance Measuring Equipment

INDEX

Made in the USA
Middletown, DE
24 July 2020